Volume 47

Advances in Genetics

Volume 47

Advances in Genetics

Edited by

Jeffrey C. Hall
Department of Biology
Brandeis University
Waltham, Massachusetts

Jay C. Dunlap
Department of Biochemistry
Dartmouth Medical School
Hanover, New Hampshire

Theodore Friedmann
Center for Molecular Genetics
University of California at San
   Diego School of Medicine
La Jolla, California

Francesco Giannelli
Division of Medical and
   Molecular Genetics
United Medical and Dental
   Schools of Guy's and
   St. Thomas' Hospitals
London Bridge, London
United Kingdom

ACADEMIC PRESS

An Elsevier Science Imprint

Amsterdam   Boston   London   New York   Oxford   Paris   San Diego
San Francisco   Singapore   Sydney   Tokyo

Academic Press
An Elsevier Science Imprint.
525 B Street, Suite 1900, San Diego, California 92101-4495, USA
http://www.academicpress.com

Academic Press
32 Jamestown Road, London NW1 7BY, UK
http://www.academicpress.com

International Standard Book Number: 0-12-017647-5

PRINTED IN THE UNITED STATES OF AMERICA
02   03   04   05   06   07   MM   9   8   7   6   5   4   3   2   1

Contents

Contributors

Numbers in parentheses indicate the pages on which the authors' contributions begin.

Peter W. Atkinson (49) Department of Entomology, University of California, Riverside, California 92521

Jean-Christophe Billeter (87) IBLS Division of Molecular Genetics, University of Glasgow, Glasgow G11 6NU, United Kingdom

Stephen F. Goodwin (87) IBLS Division of Molecular Genetics, University of Glasgow, Glasgow G11 6NU, United Kingdom

Anthony A. James (49) Department of Molecular Biology and Biochemistry, University of California, Irvine, California 92697

Andreas Keller (1) Department of Genetics and Neurobiology, Biocenter, University of Würzburg, D-97074

Jean-René Martin (1) Neural Bases of Locomotor Activity in Drosophila, NAMC, CNRS, UMR-8620, Université Paris-Sud, Centre Scientifique d'Orsay, F-91405 Orsay, France

Kevin M. C. O'Dell (87) IBLS Division of Molecular Genetics, University of Glasgow, Glasgow G11 6NU, United Kingdom

Alexandre A. Peixoto (117) Department of Biochemistry and Molecular Biology, Fundação Oswaldo Cruz, CEP 21045-900 Rio de Janeiro, Brazil

Sean T. Sweeney (1) Department of Biochemistry and Biophysics, University of California, San Francisco, San Francisco, California 94143

1

Targeted Expression of Tetanus Toxin: A New Tool to Study the Neurobiology of Behavior[*]

Jean-René Martin[†]
Neural Bases of Locomotor Activity in Drosophila
NAMC, CNRS, UMR-8620
Université Paris-Sud, Centre Scientifique d'Orsay
F-91405, Orsay, France

Andreas Keller
Department of Genetics and Neurobiology, Biocenter
University of Würzburg
D-97074 Würzburg, Germany

Sean T. Sweeney
Department of Biochemistry and Biophysics
University of California, San Francisco
San Francisco, California 94143

[*]Dedicated to the memory of Heiner Niemann (1946–1999), without whose generosity the work here described would not have been possible.
[†]To whom correspondence should be addressed.

ABSTRACT

Over the past few decades, the explosion of molecular genetic knowledge, particularly in the fruit fly *Drosophila melanogaster,* has led to the identification of a large number of genes, which, when mutated, directly or indirectly affect fly behavior. Beyond the genetic and molecular characterization of genes and their associated molecular pathways, recent advances in molecular genetics also have allowed the development of new tools dedicated more directly to the dissection of the neural bases for various behaviors. In particular, the conjunction of the development of two techniques—the enhancer-trap detection system and the targeted gene expression system, based on the yeast GAL4 transcription factor—has led to the development of the binary enhancer-trap P[GAL4] expression system, which allows the selective activation of any cloned gene in a wide variety of tissue- and cell-specific patterns. Thus, this development, in addition to allowing the anatomical characterization of neuronal circuitry, also allows, via the expression of tetanus toxin light chain (known to specifically block synaptic transmission), an investigation of the role of specific neurons in certain behaviors. Using this system of "toxigenetics," several forms of behavior—from those mediated by sensory systems, such as olfaction, mechanoreception, and vision, to those mediated by higher brain function, such as learning, memory and locomotion—have been studied. These studies aim to map neuronal circuitry underlying specific behaviors and thereby unravel relevant neurophysiological mechanisms. The advantage of this approach is that it is noninvasive and permits the investigation of behavior in the free moving animal. We review a number of behavioral studies that have successfully employed this toxigenetic approach, and we hope to persuade the reader that transgenic tetanus toxin light chain is a useful and appropriate tool for the armory of neuroethologists.   © 2002, Elsevier Science (USA).

I. INTRODUCTION

Behavior lies at the overlap of many different research fields. To some extent, behavior is the final functional goal of many systems under study, since genetic, biochemical, and physiological systems are all oriented toward providing controlled patterns of behavior, for the advantage, survival, and reproduction of the organism. Since behavior is the action of the animal in its environment and reflects the outcome of the integration and the fine tuning of all systems of the organism, it could thus possibly be considered an "emergent function," as opposed to a simple summation of its numerous single components. Nevertheless, it is obvious that a certain behavioral sequence would be limited (or constrained) by the machinery, from circuitry to molecules, subserving the performance of the living organism. Such machinery is undoubtedly (though arguably only initially) determined by genetic makeup.

Genetics and behavior have for a long time enjoyed an ambiguous relationship (Greenspan, 1995). At one end of the behavior/genetics spectrum lies the tendency of the geneticist to wish to assign each behavior to a gene, whereas at the other end of the spectrum, behavior itself is often only partially described in terms of its relative difficulties in definition, description, and quantification. For several molecular geneticists, a behavioral mutant can be considered a type of developmental mutant, and behavior is therefore considered only the secondary consequence of a defect accumulated during the development process. Conversely, for the behaviorist, what makes behavior interesting is how, when, and finally why an animal might perform a certain behavioral sequence. In these terms, we temptingly assume that "how" could refer to the description of the behavioral sequence itself, which thus must first be well described and quantified (which is not always the case, and there is much room for such effort). [For example, the design of a quantitative paradigm has made it possible to identify a highly organized fractal structure within locomotor activity (see later discussion; Martin *et al.*, in press). In addition to designing a novel courtship conditioning paradigm, Siwicki and her colleagues (McBride *et al.*, 1999) have been able to show that courtship conditioning in wild-type male flies establishes a long-term memory, lasting up to 9 days, whereas in Mushroom-Body ablated males, this memory dissipates completely within a day.] The "when" probably relates to the conjunction of the immediate external environment with the body's internal physiological needs. Finally, the "why" implies a "decision process" that most likely deals with the actual state of the organism (external environment in combination with the body's internal physiological needs) in addition to the sum of knowledge acquired from previous life experience stored somewhere in the brain, as a memory trace. Indeed, certain fly behaviors, such as courtship, have been shown to be subject to sensitization (a nonassociative form of plasticity; Kyriacou and Hall, 1984), as well as to associative modification (Siegel and Hall, 1979; Siegel *et al.*, 1984). Such results lead to the suggestion that previous experience can participate in modifying the "decision process" according to a further similarly given context, and this therefore could represent adaptative value in favor of evolutionary fitness.

Over the past few decades, the explosion of molecular genetics in general, and particularly in the fruit fly *Drosophila melanogaster*, has led to the identification of a huge number of genes, some of which, when mutated, affect various behavioral sequences directly or indirectly. However, beyond the genetic and molecular characterization of such genes and their associated molecular pathways, only very few have been revealed to be behavior-specific mutants [several previous reviews have already covered the topic of behavioral genetics (Hall and Greenspan, 1979; Hall, 1982, 1994a; Heisenberg, 1994; Kyriacou and Hall, 1994; Pflugfelder, 1998)]. On the other hand, recent advances in molecular genetics have contributed to the development of new tools dedicated more specifically to the dissection of the neural bases of behaviors. In particular, the conjunction of the development of two techniques, the enhancer-trap detection system (O'Kane and Gehring, 1987) and the targeted gene expression system, based on the yeast transcriptional activator GAL4, has led to the binary enhancer-trap P[GAL4] expression system, which allows the selective activation of any cloned gene in a wide variety of tissue- and cell-specific patterns (Brand and Perrimon, 1993). Initially, this system allows the anatomical characterization of specific neuronal circuitry, by way of the expression of a variety of marker reporter genes. Similarly, this genetic tool also allows the expression of toxic genes, in particular the tetanus toxin light chain (TeTxLC) gene, expression of which is found to specifically block synaptic transmission, leading to neuronal silencing. The ability to thus silence neurons is of particular facility to the study of neuronal circuitry in an otherwise normally functioning fly.

This review focuses on the use of transgenic TeTxLC to dissect the neurobiology of behavior. We will describe how and why this tool has been constructed, its mechanism of action, and the enhancer-trap P[GAL4] system used to specifically target it in precise groups of neurons. We will then describe how certain behavioral sequences have been impaired in direct correlation with relevant electrophysiological data. Finally, we will present results of research in which this toxigenetic tool has been successfully used to study various sensory functions in addition to dissecting some of the complexity of central brain function.

II. MECHANISM AND ACTION OF TETANUS TOXIN LIGHT CHAIN

A. An introduction to the clostridial toxins

Tetanus toxin, the causative agent of tetanus (or "lockjaw"), is produced by the gram-positive anaerobe *Clostridium tetani* and belongs to the family of clostridial neurotoxins that normally are found as a dichain holotoxin consisting of a heavy chain (H chain, ~100 kDa) and a light chain (L chain, ~50 kDa; for a review of the clostridial toxins see Schiavo *et al.*, 2000). Clostridial neurotoxins inhibit neurotransmitter release in a neuron when intracellularized, resulting in paralysis of the infected organism. For all the clostridial neurotoxins, when applied extracellularly, the L chain cannot cause paralysis when isolated from the H chain (Poulain *et al.*, 1991; Mochida *et al.*, 1989). It is the L chain of the toxin that contains the intracellularly toxic moiety that is responsible for blocking synaptic exocytosis when delivered into the cytoplasm of a neuronal cell (Li et al., 1994; Mochida et al., 1989, 1990). Delivery of the tetanus toxin light chain (TeTxLC) into neuronal cells (such as the buccal ganglia of the mollusk Aplysia californica), either as purified protein (Mochida et al., 1989) or as mRNA encoding the L chain (Mochida et al., 1990), is sufficient by itself to inhibit exocytotic neurotransmitter release. The L chain contains a sequence motif, His-Glu-X-X-His, found in a variety of Zn^{2+}-dependent metalloproteases (Jongeneel et al., 1989; Fujii et al., 1992). In such metalloproteases, histidines in the motif are involved in the coordination of a Zn^{2+} ion that is essential for catalytic activity. In TeTxLC, this motif has been shown to bind a single atom of Zn^{2+} per molecule of L chain, suggesting that the L chain of tetanus toxin is similar to that of thermolysin-type metalloproteases (Schiavo et al., 1992a). The proteolytic substrate of TeTxLC was identified by two groups simultaneously as synaptobrevin (syb), also known (originally) as VAMP (vesicle-associated membrane protein; Schiavo et al., 1992b; Link et al., 1992).

Synaptobrevin is a small C-terminally anchored membrane protein of 18 kDa that was first identified as a component of synaptic vesicles in Torpedo electric organ (Trimble et al., 1988) and later cloned (and in the process, renamed) from a variety of species from yeast to human (Gerst et al., 1992; Protopopov et al., 1993; Yamasaki et al., 1994a; Hunt et al., 1994; Baumert et al., 1989; Elferink et al., 1989; Archer et al., 1990), including two forms from Drosophila (Südhof et al., 1989; Chin et al., 1993; DiAntonio et al., 1993). Synaptobrevin has been found to interact with a number of presynaptic proteins of importance to the process of regulated exocytosis (for a detailed review, see Schiavo et al., 2000). However, the central importance of synaptobrevin to the process of neurotransmitter release is evidenced by the observation that cleavage of synaptobrevin by TeTxLC elicits a complete blockade in evoked exocytosis (Schiavo et al., 1992b; Link et al., 1992). The identification of synaptobrevin as the substrate for the L chain of tetanus toxin and five of the botulinal toxins (O'Kane et al., 1999) indicates a core role for synaptobrevin in the process of synaptic-vesicle exocytosis. Synaptobrevin is found in a tight complex with two other synaptic proteins, syntaxin and SNAP-25, both of which are also targets for three clostridial neurotoxins; loss of either, again, results in failure of evoked synaptic exocytosis. Two forms of synaptobrevin have been found in Drosophila, a ubiquitously expressed, TeTxLC-insensitive form, dsyb (Südhof et al., 1989; Chin et al., 1993; Sweeney et al., 1995), and a neuronally expressed, TeTxLC-sensitive form, dn-syb (DiAntonio et al., 1993; Sweeney et al., 1995). These facts allied to the apparent specificity and cell autonomy of TeTxLC have made it an attractive tool for studying exocytosis in a number of preparations including Drosophila. Linking these properties to a powerful transgenic targeted expression system has additionally made TeTxLC a particularly useful tool for the dissection of behaviors in Drosophila.

B. The P[GAL4] system coupled to tetanus toxin light chain

The ability to target the expression of a particular gene to specific cell types in a metazoan is a very powerful approach to the study of development and function of cells or groups of cells. This approach may be useful in two ways: First, the role of a gene product itself may be studied; and second, insights into the developmental role, interactions and functions of cells under study may be gained. The P[GAL4] system developed by Brand and Perrimon (1993) is a method devised for *Drosophila* that can be used to target gene expression in specific cell types or tissues. A brief description of the technique is outlined here, but for reviews on the system and its use, the reader is referred to van Roessel and Brand (2000) and Brand and Dormund (1995). The P[GAL4] system can employ cloned promoter sequences or "enhancer trapping" to direct gene expression. Enhancer trapping employs a weak promoter coupled to a reporter (or in the P[GAL4] system, a transcriptional activator) that cannot normally drive expression in the absence of a transcriptional enhancer (O'Kane and Gehring, 1987). Insertion of this construct into the genome close to endogenous transcriptional enhancers activates transcription of the transgene (Bellen *et al.*, 1989; Wilson *et al.*, 1989; Bier *et al.*, 1989). Transcriptional promoters and enhancers dictate temporal and spatial patterns of gene expression, and by thus hijacking this mechanism, GAL4 can be expressed in a pattern of choice. The generation of a large number of possible GAL4 expression patterns is then made possible by mobilizing the P[GAL4] element in a classic P-element-mediated mutagenesis (Robertson *et al.*, 1988; Cooley *et al.*, 1988). In the second component part of the system, a P-element vector containing five tandemly arrayed GAL4 binding sites (UAS_G) [the use of five sites has been shown to optimize transcriptional activation of the target gene] sequences upstream of the gene that the experimenter wishes to ectopically express is constructed (in this case the gene for TeTxLC) in a separate strain of flies (Brand and Perrimon, 1993).

To express TeTxLC in a desired pattern requires the crossing of the two strains, and the expression of TeTxLC will occur in the progeny only in cells where GAL4 is expressed (Figure 1.1, see color insert). Using this system, the ectopically expressed gene product can be toxic, since it is only in a cross with a P[GAL4] containing line that the gene will be expressed. Crosses can be carried out in large numbers and repeated ad infinitum to prepare flies expressing TeTxLC in a pattern of interest. The P[GAL4] system has been used in a number of studies to neuronally express other gene products in order to study the role of certain genes, or the role of the cells in which they have been expressed, in fly behavior (Ferveur *et al.*, 1995, 1997; O'Dell *et al.*, 1995; Connolly *et al.*, 1996; McNabb *et al.*, 1997; Renn *et al.*, 1999a; Zars *et al.*, 2000a,b; Gatti *et al.*, 2000). In some cases, the P[GAL4] system has been used to express cell-death genes in order to ablate brain structures thought to be mediating the behavioral response under study (McNabb *et al.*, 1997; Renn *et al.*, 1999a). However, for behavioral studies, ablation may not be the ideal weapon of choice given that ablation of a neuron may affect the development or function of surrounding or supported neurons (see Lin *et al.*, 1995). Bringing TeTxLC under control of the P[GAL4] system was originally envisioned as a means of circumventing such problems. It was originally envisioned that by specifically targeting and blocking only the release of neurotransmitter, the developmental problems of ablation could be avoided, and the role of a specified neuron or group of neurons within the context of behavior focused upon. However, as described below, in some cases expression of the toxin in neurons may cause subtle neuroanatomical defects, possibly related to activity-dependent development that may impinge on behavior.

C. The action of tetanus toxin light chain in *Drosophila*

1. Effects in nonneuronal tissues

A set of transgenic flies bearing UAS-TeTxLC have been made [using a form of the TeTxLC gene that has been resynthesized to aid eukaryotic expression by replacing the adenine/thymine nucleotide bias of the codon usage of the original gene (Eisel *et al.*, 1993)], and the effects of expression of TeTxLC in various tissues of the fly tested (Sweeney *et al.*, 1995). First, driving expression of TeTxLC in the muscle from embryogenesis to adulthood causes no observable developmental defect, and indeed the flies expressing the toxin in this manner mature to adulthood (Sweeney *et al.*, 1995; Davis *et al.*, 1998). Furthermore, in an electrophysiological assay of neuromuscular junction (NMJ) function, postsynaptic expression of TeTxLC is seen to have no effect on excitatory junction currents (Sweeney *et al.*, 1995). This result also serves to demonstrate that there is no retrograde transport of the light chain from the muscle to the synapse. Indeed, expression of TeTxLC in larval salivary glands, a tissue with a high secretory activity, has no effect on larval viability (Sweeney, 1996). Thus, we suspect that outside of the nervous system, TeTxLC has no target of obvious physiological importance in nonneuronal cells, a result that essentially allows the experimenter to ignore extraneous expression of TeTxLC in nonneuronal tissue during behavioral experiments.

2. Effects in neuronal tissues

In embryos expressing TeTxLC throughout the nervous system, no morphological defects are apparent (Sweeney *et al.*, 1995). Using an antibody raised to human synaptobrevin to assay for the presence of dn-syb at the embryonic NMJ, dn-syb immunoreactivity is totally abolished in the presence of TeTxLC (Sweeney *et al.*, 1995). As would be expected for a panneural removal of dn-syb, embryos in which TeTxLC is expressed thoughout the nervous system are almost totally paralyzed and show no coordinated movement at the end of embryogenesis, a developmental time point when the characteristic peristaltic movement of larval behavior becomes obvious (Sweeney *et al.*, 1995; Baines and Bate, 1998). Nonetheless, synapses form normally on the muscle, and iontophoretic application of glutamate to the synapse results in an otherwise normal depolarization of the muscle (Sweeney *et al.*, 1995). Examining the action of the neuromuscular synapse by stimulating the nerves innervating abdominal body wall muscles via a suction electrode and recording currents in the muscle demonstrates that synapses expressing TeTxLC are unable to elicit any excitatory junction currents (EJCs), and that the miniature excitatory junction currents (mEJCs) are reduced in number by 50% (Sweeney *et al.*, 1995).

This suggests that the embryonic nervous system assembles correctly in the absence of evoked synaptic release. Previous work on the development of the embryonic NMJ showed that contact between the motor neurons and their target muscles induces the clustering of glutamate receptors in the postsynaptic membrane (Broadie and Bate, 1993; Nishikawa and Kidokoro, 1995; Saitoe *et al.*, 1997) and that this clustering is dependent on presynaptic electrical activity (Broadie and Bate, 1993). The ability of the postsynaptic architecture to assemble correctly in the absence of evoked release (Sweeney *et al.*, 1995, Broadie *et al.*, 1995, Deitcher *et al.*, 1998) therefore remains a paradox. In mammalian systems, postsynaptic organization is largely achieved in the absence of activity (through the activities of molecules such as agrin and rapsyn) though the maintenance of the synapse is generally dependent on activity (Hall and Sanes, 1993). Studies of mice defective for a protein of central importance to synaptic transmission, Munc-18, have shown that a mammalian brain can also assemble in the absence of evoked release (Verhage *et al.*, 2000).

D. Specificity of the toxin

Two forms of synaptobrevin are present in *Drosophila*, a predominantly neuronally expressed form, dn-syb (DiAntonio *et al.*, 1993), and a more ubiquitously expressed form, dsyb (Chin *et al.*, 1993). dsyb exists in two splice forms, dsyb-a and dsyb-b, the latter bearing an additional 10 amino acids on the C-terminal intravesicular tail, the function of which is unclear (Chin *et al.*, 1993). Mammalian syb is cleaved by TeTxLC between amino acids Q76 and F77, adjacent to the cytoplasmic side of the transmembrane domain (Schiavo *et al.*, 1992b). Both dsyb and dn-syb bear the QF motif in a similar position, initially indicating that TeTxLC might cleave both forms. However, using dsyb and dn-syb protein translated *in vitro*, dsyb was found to be insensitive to the protease action of TeTxLC, whereas dn-syb was readily cleaved, a finding that was later supported by *in vivo* expression of the toxin (Sweeney *et al.*, 1995).

The observation that dsyb is not cleaved by TeTxLC is surprising. Some studies have shown that the sequence of the cleavage site can have an influence on the activity of the toxin on its substrate (Yamasaki *et al.*, 1994b; Regazzi *et al.*, 1996). However, it has become clear from a number of substrate-specificity studies that the structure of synaptobrevins may feature more predominantly in the ability of the toxin to recognize and cleave its substrate (reviewed in O'Kane *et al.*, 1999). The importance of the insensitivity of dsyb to cleavage by TeTxLC is demonstrated by parallel TeTxLC expression studies in the mouse. A ubiquitously expressed mammalian homolog of synaptobrevin, cellubrevin, is a target for TeTxLC (McMahon *et al.*, 1993), and cleavage of cellubrevin may account for the severe developmental defects observed when TeTxLC is expressed in the mouse seminiferous epithelium (Eisel *et al.*, 1993). Cellubrevin has been found to be involved in a more general exocytotic process (Galli *et al.*, 1994; Breton *et al.*, 2000), as dsyb may be, since embryos mutant for dsyb are found to be defective for outgrowth of tracheal branches (Jarecki *et al.*, 1999), a phenotype not observed in TeTxLC-expressing embryos. Since dsyb is the protein most closely related to dn-syb in *Drosophila*, and no other homologs appear in the *Drosophila* genome that resemble targets for TeTxLC (Lloyd *et al.*, 2000), the insensitivity of dsyb to toxin cleavage may indicate that dn-syb is likely to be the only substrate of the metalloprotease activity of TeTxLC in flies.

Two proposed activities of TeTxLC in addition to the metalloprotease activity have been described. Additional activities of the toxin so far identified have their bearing on synaptic exocytosis and are independent of metalloprotease activity and cell autonomous. (These other proposed activities are contentious and are reviewed and assessed in detail in O'Kane *et al.*, 1999.) Mutants of TeTxLC that lack the metalloprotease activity, when expressed in neuronal tissue in *Drosophila*, appear to cause no defects in neuronal function or activity (Sweeney *et al.*, 1995). Similarly, expressing the mutant forms of TeTxLC does not appear to affect the level or distribution of dn-syb protein at the neuromuscular junction (Sweeney, 1996), and embryos expressing mutant TeTxLC throughout the nervous system are indistinguishable from wild-type embryos in their motility, hatching, viability, or behavior (Sweeney *et al.*, 1995; Heimbeck *et al.*, 1999; Kaneko *et al.*, 2000). Thus, if cessation of synaptic transmission is the desired effect of TeTxLC expression, principally as a tool for dissecting behavior in *Drosophila*, then use of the inactive light-chain mutants as a control is advisable in order to discount any potential additional activity of the toxin. Nonetheless, the possibility of other substrates for TeTxLC becomes an issue only when a particular role is ascribed to dn-syb with transgenic TeTxLC. In exploring the role of dn-syb in the development of the photoreceptors, Hiesinger *et al.* (1999) examined eyes lacking dn-syb by expressing TeTxLC in the fly eye. By also examining eyes genetically deficient in clonal patches for dn-syb, potential additional activities of TeTxLC were efficiently controlled for and the effects of TeTxLC expression could be attributed solely to the loss of dn-syb function by comparison of the two conditions.

From the number of studies performed so far using UAS-TeTxLC, it is clear that TeTxLC blocks exocytosis of classical small neurotransmitters by small synaptic vesicles (SSVs) (Sweeney *et al.*, 1995; Broadie *et al.*, 1995). Whether TeTxLC also blocks exocytosis of neuropeptides from large dense core vesicles

(LDCVs) in flies remains a moot point. LDCV-mediated secretion is certainly blocked by TeTxLC in mammalian cells (Penner *et al.*, 1986; Ahnert-Hilger *et al.*, 1989). However, the process of LDCV-mediated secretion and the molecules that might underpin this process are relatively understudied subjects in *Drosophila*. Use of TeTxLC to block secretion of neuropeptides involved in identified behaviors has been inconclusive in this regard. First, in the study by McNabb *et al.* (1997) on the role of eclosion hormone (EH)-containing cells in eclosion behaviors, ablation of the EH-containing cells by expression of cell-death genes altered eclosion behavior. Expression of TeTxLC in the same cells had no apparent effect (Sue McNabb, pers. comm.). In a second study by Kaneko *et al.* (2000), expression of TeTxLC in cells expressing a neuropeptide known to mediate circadian rhythms (Renn *et al.*, 1999a) caused minor defects in behavioral rhythms. The decrement in behavior was nowhere near as severe, however, as the genetic knockout of the gene encoding the peptide or *pdf-gal4*-mediated ablation of the neurons that contain it (Renn *et al.*, 1999a).

One synapse type not amenable to study by the expression of TeTxLC would be an "electrical" or gap-junction synapsed connection. Gap-junction synapses are usually employed to relay "fast" information and in order to relay such information are directly electrically coupled. The giant descending neuron (GDN) of *Drosophila*, which mediates the fast component of the escape reflex, is coupled via gap junctions to the tergotrochanter motor neurons (TTMn) and the peripherally synapsing interneuron (PSI; King and Wyman, 1980). Oddly, this synapse is of a "mixed" nature, comprising gap junctions and a chemical component (Blagburn *et al.*, 1999). Allen *et al.* (1999) expressed TeTxLC in the GDN and studied the consequence of this expression electrophysiologically, finding that the initial firing of the GDN to TTMn synapse was unaffected by the toxin, but subsequent firing of the synapse was impaired. Thus, TeTxLC might lend insight into the operation of this atypical synapse.

E. Developmental consequences of tetanus toxin light-chain expression

On closer inspection, expression of TeTxLC in some instances does appear to have developmental consequences, though these appear initially to be due to loss of activity-dependent mechanisms. Baines *et al.* (1999) examined, *in situ*, the development of synaptic input patterns onto motor neurons while expressing TeTxLC in the same motor neurons in order to observe the effects that the loss of evoked synaptic activity might have on these developing cells. Expression of TeTxLC in the particular motor neurons examined caused the failure of synaptic transmission between these neurons and their paired muscles. In addition to this expected result, electron microscopy and recordings from the soma of the motor neurons demonstrated that, in turn, the motor neurons themselves receive considerably less synaptic input when their output is blocked (Baines *et al.*, 1999).

This reduction in synaptic input in the absence of output may be a consequence of a blocking of a retrograde signal from the postsynaptic neuron. Another related mechanism may be that the neuronal circuits of the embryo may require activity to refine their normal synaptic input complement and by blocking activity of the postsynaptic cell, a reduction in the normal synaptic input may be produced by competitive mechanisms.

When expressing TeTxLC in larval chemosensory neurons, Heimbeck *et al.* (1999) found that chemosensory cells expressing the toxin exhibited weak anatomical differences in comparison to wild-type neurons. At the *Drosophila* larval neuromuscular junction, synaptic activity dramatically affects the growth and expansion of the synapse (Ganetzky and Wu, 1993; Budnik *et al.*, 1990; Kernan *et al.*, 1991), and it is likely that the morphological defects observed in TeTxLC-expressing sensory neurons reflect such activity-dependent changes. In the same way, Hiesinger *et al.* (1999) have studied the influence of a lack of functional dn-syb during optic-lobe development, by comparing the targeted expression of the TeTxLC and dn-syb null mutant eye mosaics. Both methods led to similar disturbances in the columnar organization of visual neuropils and photoreceptor terminal neuropils, suggesting that the effects of the TeTxLC are mediated by the block of the activity of the dn-syb, which in turn regulates cell adhesion molecules.

III. FROM ELECTROPHYSIOLOGY TO BEHAVIOR

One of the greatest difficulties in developing new tools for investigating behavior is devising a test to ascertain that the tool will work efficiently *in vivo*. As has been well demonstrated by Sweeney *et al.* (1995), expression of TeTxLC in the presynaptic neuron eliminates excitation-evoked synaptic transmission, as revealed by electrophysiological recording of excitatory junction currents at an identified neuromuscular junction. Further, strong proof of the efficacy of the technique would be to have a system where a direct and causal link can be drawn between electrophysiology and a related behavior. Such a link has now been well demonstrated in adult flies using the leg resistance reflex (Reddy *et al.*, 1997). Briefly, in *Drosophila*, as in larger insects, when the tibio-femoral joint is flexed passively, the femoral chordotonal organ is activated, and this in turn excites two motor neurons innervating the tibial extensor muscle. Activation of this muscle resists the movement; hence, it is called a "resistance reflex" (Bassler, 1993; Burrows, 1987). Reddy *et al.* (1997) have elegantly demonstrated by extracellular recordings from the extensor muscles of the tibia that the mutation $Glued^l$ (Gl^l) alters the anatomy of the axons of the femoral chordotonal organ and consequently disrupts the resistance reflex between the sensory neurons of the chordotonal organs and the tibial extensor motor neurons. By expressing the tetanus toxin in the sensory neurons, via the use of two independent enhancer-trap P[GAL4] lines (P[GAL4]50y and P[GAL4]c362), Reddy and co-workers were able to eliminate the resistance reflex, thus creating a phenocopy of the defective reflex observed in Gl^l heterozygotes. This work represents the first demonstration of a direct correlation between blocked synaptic transmission as assayed electrophysiologically and the blocked transmission leading to a disruption of the associated behavior. This set of experiments therefore contributes an initial validation of the use of the targeted expression of TeTxLC to block neurons and subsequently dissect behavior.

IV. DISSECTING BEHAVIOR

A. Sensory systems

Sensory modalities are of principal importance in informing and guiding the animal in an environment and for the correct and appropriate performance of tasks that might ensure survival and reproduction. Olfaction, mechanoreception, taste, and vision are major modalities currently under study that are likely to inform us how an animal captures and responds to environmental cues.

The sensory structures of the *Drosophila* nervous system have already been studied extensively using genetic, behavioral, and physiological techniques (for reviews of visual mutants, see Pflugfelder, 1999, and Heisenberg and Wolf, 1984; for review of olfactory mutants, see Carlson, 1996). The anatomy of the sensory organs and that of the connected neuropils have been well established (Fischbach and Dittrich, 1989; Stocker, 1994); and, unlike the central brain, sensory structures are accessible to electrophysiological methods. Electroretinograms (Coombe, 1986) as well as electroantennograms (Alcorta, 1991) have been used to analyze behavioral mutants. For these reasons, the sensory systems of *Drosophila* are amenable to the toxigenetic method, allowing the analysis of relatively simple neural circuits underlying early processing of sensory inputs.

1. Mechanoreception

The main external stimuli known to trigger behavioral responses in *Drosophila* are humidity, temperature, mechanical stimulation, taste, odor, and light. The hygroreceptors (sensing humidity) in *Drosophila* are housed in the arista, whereas thermoreceptors are located in the distal antenna (3rd antennal segment) (Sayeed and Benzer, 1996). The stimuli relayed by the mechanosensory system include touch, relative body-part position (proprioception), and sound. These stimuli are mediated by neurons with single dendrites (type I mechanoreceptors) in sensory bristles, chordotonal organs, campaniform sensilla, and multidendritic type II mechanoreceptors (Jan and Jan, 1993). Chordotonal organs and type II mechanoreceptors are internal sensory organs, whereas sensory bristles and campaniform sensilla use external sensory structures that detect external mechanical signals. Sensory bristles are numerous and distributed over the entire body surface of *Drosophila*. Campaniform sensilla, in contrast, are restricted to the wings, the basal and middle segment of the halteres, and the trochanter and femur of the legs (Bryant, 1978). For sound perception, the Johnston's organ is used. It is derived from several chordotonal organs and is located in the second antennal segment. The Johnston's organ is stimulated by vibrations of the funiculus and arista, the latter acting as a "sail" (Manning, 1967).

As described earlier, Reddy *et al.* (1997) used two P[GAL4] lines to drive TeTxLC in sensory neurons of the femoral chordotonal organ and thereby blocked the resistance reflex. This reflex normally resists passive flexing of the tibio-femoral joint by activating the tibial extensor muscle. A logical next step might be to use these flies to study the role of the resistance reflex in the coordination of legs during walking. In *Drosophila*, leg coordination in straight walking and turning has been described (Strauss and Heisenberg, 1990). Parameters such as the mean step length, step frequency, and mean recovery stroke duration could thus be measured in flies with blocked chordotonal organs and compared with the known wild-type values. A P[GAL4] line expressing in yet another type of mechanoreceptor, the campaniform sensilla in the halteres, is now available (Keller *et al.*, in press). The campaniform sensilla in the wing blade of *Calliphora* are speculated to monitor the passage of a deformational wave that travels along the wing during each wing beat (Dickinson, 1990), whereas the campaniform sensilla in the halteres are equilibrium organs sensing Coriolis forces resulting from rotations of the body and mediating corrective reflexes (Nalbach and Hengstenberg, 1994). Using the P[GAL4] line expressing in the campaniform sensilla in the halteres, it should be possible to analyze the contribution of these campaniform sensilla to course control during flight. Further, the use of TeTxLC expression in synapses of *Drosophila* flight circuits may help to elucidate the relative contribution of the electrical and chemical components of this circuitry. Previous studies (Trimarchi and Murphey, 1997) have shown that the synapses made between the haltere afferents and a flight motor neuron have both an electrical and a chemical component. In insects, innexins are thought to be the counterparts of the gap-junction-forming connexins (Phelan *et al.*, 1998). Mutations in one of the *Drosophila* innexin genes were shown to disrupt the flight circuit, although not entirely, as a cholinergic component remained (Trimarchi and Murphey, 1997). The use of this mutational analysis combined with targeted TeTxLC expression may help to elucidate the relative contributions of each component to this synapse.

2. Smell and taste

The structures processing chemosensory (olfactory and gustatory) information in *Drosophila* are organized differently in adults and larvae. In the adult, olfactory and gustatory information is processed at different successive levels. Chemosensory neurons are located on the third antennal segment, the maxillary palps, the labellum, the pharynx, the tarsi, the wings, and the female genitalia (for a review of chemoreception, see Stocker, 1994). The olfactory receptors of the maxillary palps and the third antennal segment (plus certain pharyngeal sensilla) project to the antennal lobe. The antennal lobe is a conspicuous brain structure composed of 43 anatomical subunits, referred to as glomeruli (Laissue *et al.*, 1999). Interestingly, individual afferent chemoreceptor fibers are invariably glomerulus-specific (Stocker *et al.*, 1983). The anatomical organization of the antennal lobes led to the speculation that individual glomeruli are functionally specialized. This idea is further supported by the finding that different odors excite specific subsets of glomeruli (Rodrigues, 1988). The major targets of the olfactory receptors in the glomeruli are local neurons [LNs, also known as local interneurons (LocI)] and projection neurons [PNs, synonymous with relay interneurons (RI) in *Manduca* and *Schistocera* (Christensen *et al.*, 1993; MacLeod and Laurent, 1996)]. PNs link single or few glomeruli with the mushroom-body calyx and the lateral protocerebrum.

In a screen for P[GAL4] lines with specific expression in the larval and adult chemosensory systems, Heimbeck *et al.* (1999) have found enhancer-trap lines well suited to structure function mapping in the olfactory pathway (Stocker *et al.*, 1997). In this study, a P[GAL4] line GH298 was found to label olfactory local neurons that arborize in most or all antennal lobe glomeruli. LNs have been shown to inhibit projection neurons. In *Schistocera*, this LN-mediated inhibition of PNs underlies the synchronization of PN ensembles (MacLeod and Laurent, 1996) for which a role in odor coding has been postulated. If a similar mechanism is used in *Drosophila*, GH298-driven TeTxLC expression in LNs would allow the desynchronization of PNs and would represent a test of the behavioral relevance of such synchronization in, for example, odor recognition or discrimination as well as olfactory learning.

The paths olfactory information takes to reach the integrative structures of the central brain are the projection neurons (PNs). PNs connect glomeruli with the mushroom-body calyx and the lateral protocerebrum. The P[GAL4] line GH146 drives expression in about 100 projection neurons (Stocker *et al.*, 1997). In addition, most but not all glomeruli are stained. This P[GAL4] line offers the opportunity to block the connections of a distinct set of glomeruli with the central brain by expressing TeTxLC. So far, no such studies have been performed in adult flies to directly investigate olfactory recognition and learning. However, the effects of the blockade of such neurons are under investigation in another behavioral test, the courtship paradigm (see later discussion).

To date, a directed disturbance of the olfactory circuitry at the primary level (e.g., sensory neurons) has not been reported. However, J.-M. Devaud and A. Ferrus (pers. comm.) have observed that expressing TeTxLC in various subsets of olfactory sensory neurons modifies the behavioral response (attraction or repulsion). For example, blocking specific olfactory inputs produces shifts from avoidance to attraction toward certain odorants at high concentration. In the near future, a set of enhancer-trap P[GAL4] lines with complementary and/or overlapping expression patterns might be used to dissect the neuronal combinatory code for odor recognition.

Drosophila larvae respond to olfactory and gustatory stimuli. Larvae have three major chemosensory structures on the head: the dorsal organ, consisting of the perforated dome and its six encircling receptors; the terminal organ; and the ventral organ (Kankel *et al.*, 1980). Heimbeck *et al.* (1999) have studied chemically induced behavior with the P[GAL4] line GH86, which drives expression in the dorsal organ, the terminal organ, and internal chemosensory cells of the mouth parts. Expression in both the dorsal organ and the terminal organ in this P[GAL4] line is restricted to a set of 16–18 neurons. Most or all dendrites of the labeled dorsal organ neurons are located in the perforated central dome, usually considered the only site of olfaction in larvae (Stocker, 1994). Expression does not include all dorsal organ sensory neurons. This P[GAL4] line allowed Heimbeck and co-workers to test the chemospecificity of the labeled receptors. Driving TeTxLC expression with P[GAL4] line GH86 produced larvae anosmic for butanol, ethyl acetate, *n*-octyl acetate, and propionic acid, indicating that the olfactory receptors for these chemicals are silenced. In contrast, cyclohexanone still elicited a response, though this response was reduced compared with the parental lines, making it likely that some (but not all) receptors sensitive to cyclohexanone were blocked. In gustatory choice tests, the responses to fructose, sucrose, and sodium chloride were reduced compared with the parental lines. This reduction may be an effect of the blocking of receptors in the terminal organ and/or pharynx. However, an involvement of the dorsal organ receptors cannot be ruled out. The remaining response is likely to be mediated by the unlabeled cells of the terminal organ, the ventral organ, and the pharyngeal and epidermal sensilla.

3. Vision

Photoreception in flies is mediated by the two compound eyes, the three dorsal ocelli, extraretinal photoreceptors at the compound eye's posterior margin (Yasuyama and Meinertzhagen, 1999), and brain neurons containing an identified deep brain photoreceptor (Emery *et al.*, 2000). The information gathered by the compound eyes is processed in the optic lobes, the most prominent structures of the adult brain (for a detailed description of optic lobe neuroanatomy, see Fischbach and Dittrich, 1989; for *Musca*, see Strausfeld, 1976). The optic lobes consist of four neurocrystalline neuropils (lamina, medulla, lobula, and lobula plate) through which information from single visual elements (consisting of photoreceptors directed to the same point in space) is processed in columns (Figure 1.2). All through the optic lobe there is crosstalk between neighboring columns (for a complete catalog of synaptic connections in the lamina, see Meinertzhagen and O'Neil, 1991). The optic-lobe neuroarchitecture as well as the repertoire of visually guided behaviors in *Drosophila* has been extensively studied. Mutants with known molecular or structural defects have occasionally been used (reviewed in Pflugfelder, 1999; Heisenberg and Wolf, 1984). In *Musca* and *Calliphora* flies, some large neurons are accessible by electrophysiology,

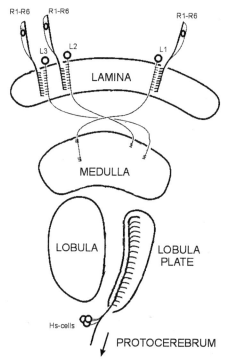

Figure 1.2. Schematic view of an optic lobe (horizontal section, anterior to the left; modified after Douglass and Strausfeld, 1996) showing the three parallel pathways between the lamina and medulla and one class of tangential large-field neurons projecting from the lobula plate to the protocerebrum. The columnar lamina monopolar cells L1 to L3 receive the same input from the R1–6 photoreceptors and terminate at characteristic levels in the medulla. The functional relevance of this parallel processing of visual information is not known. As an example for tangential large-field neurons, the three HS cells with dendritic arborizations in the anterior part of the lobula plate are shown. The seven VS cells with their arborizations at the posterior surface of the lobula plate are not shown for clarity.

and the cellular repertoire of the optic lobes has been shown to be similar to that in *Drosophila* (Fischbach and Dittrich, 1989). This makes the optic lobe of *Drosophila* a tempting structure in which to analyze information processing using toxigenetics. So far the effects of TeTxLC expression in the visual system have been studied only during development by Hiesinger *et al.* (1999), using two different P[GAL4] lines. First they expressed TeTxLC in photoreceptors using the P[GAL4] line GMR-GAL4, which drives expression in all photoreceptor neurons throughout development under control of the *glass* DNA-binding site (Moses and Rubin, 1991; Freeman, 1996). In addition, they used a less specific P[GAL4] line, Mz1369, that drives expression in many cell types of the visual system throughout the pupal period. TeTxLC expression during development disturbs the columnar organization of visual neuropils as well as the morphology of photoreceptor terminals. In addition, the IrreC-rst protein (a cell adhesion molecule of the Ig superfamily) was increased in neuropils after Mz1369-driven TeTxLC expression. After GMR-GAL4-driven expression, immunoreactivity against two other cell adhesion molecules but not against IrreC-rst was increased. Unfortunately, flies expressing TeTxLC in a Mz1369-dependent way are not useful for structure–function correlation because in the adult fly expression is widespread in the central brain. Flies expressing TeTxLC in all photoreceptors using GMR-GAL4 prove to be blind, showing that TeTxLC is capable of completely blocking synaptic transmission in photoreceptor cells (Keller and Heisenberg, 1998).

Postsynaptic to the photoreceptors are the lamina monopolar cells L1 to L3. Ever since the publication of the first comprehensive neuroanatomical study of the lamina neurons (Cajal and Sanchez, 1915), the functional relevance of parallel processing in the lamina has been debated. The *Drosophila* structural mutant *vacuolar medulla* (*vam*) shows age-related degeneration of the lamina monopolar cells L1 and L2 accompanied by a loss of optomotor responses (Coombe and Heisenberg, 1986). Whether the degeneration in the *vam* mutant is confined to only L1 and L2 is uncertain, but at least the only other monopolar cell postsynaptic to photoreceptors, L3, is intact. This experimental study was the first direct evidence that L1 and/or L2 is necessary for optomotor responses. It is now possible to study information processing in the lamina in more detail using appropriate P[GAL4] lines to drive TeTxLC expression. Thus far two P[GAL4] lines expressing TeTxLC in lamina monopolar cells have been analyzed (Keller *et al.*, 1999). If TeTxLC expression is driven in L2 neurons only, flies still show optomotor responses, albeit at a reduced level. Flies expressing TeTxLC with another P[GAL4] line in yet-unidentified lamina monopolar cells show no optomotor responses. These flies can, however, still orient toward large black stripes, similarly to *vam* mutant flies. A more detailed toxigenetic behavioral analysis of these two lines should help elucidate the functional relevance of parallel processing in the lamina.

Another question that may be amenable to the GAL4/TeTxLC intervention technique is the functional role of the different types of large-field tangential neurons originating from the lobula plate and terminating either in the contralateral lobula plate or in the protocerebrum. In an ongoing screen, GAL4 lines that label 5 types of lobula plate large-field tangential neurons have been identified in *Drosophila*. Two of the 5 types terminate in the contralateral lobula plate, whereas the others terminate in the posterior slope (2 types) or the ventrolateral protocerebrum (1 type) (Otsuna and Ito, 2000). In *Calliphora*, 21 classes of large-field tangential neurons, which differ in anatomical and electrophysiological properties, have been identified in the lobula plate (Hausen, 1981). Again, a *Drosophila* structural mutant has helped to attach certain functions to identified neurons. In the mutant *optomotor blind* (omb^{H31}), two classes of lobula plate tangential neurons are missing, the VS and HS cells (Heisenberg *et al.*, 1978). Detailed behavioral analysis showed that in omb^{H31} flies the horizontal large field response, evoked by a rotating panorama, is missing. However, they still display the so-called object response that is evoked by small objects moving from front to back (Bausenwein *et al.*, 1986). Remarkably, in other visually guided behaviors such as motion-induced landing and lift/thrust responses, omb^{H31} flies were found to be normal (reviewed in Pflugfelder and Heisenberg, 1995). This led to the hypothesis that the horizontal large-field response is mediated by the HS cells, whereas the object response is mediated by another class of large-field tangential neurons. This was confirmed by microsurgical experiments in *Calliphora* (Hausen and Wehrhahn, 1990). In omb^{H31}, most of the 3′-regulatory region of the *omb* gene is removed (Pflugfelder *et al.*, 1990). In a mutant in which only the distal one-third of this region is deleted, the HS and VS cells are still missing, but their optomotor defects are less severe (Brunner *et al.*, 1992). Apparently, structural defects other than the lack of HS and VS cells interfere with optomotor responses. Six additional structural defects in omb^{H31} are described (Pflugfelder and Heisenberg, 1995), but to what extent these defects are related to optomotor behavior is not clear. On the other hand, some optomotor responses can be performed without HS and VS cells. Since VS-cell-specific P[GAL4] lines are now available (Kerscher *et al.*, 1995), it is possible to specifically block this class of lobula plate neurons and study their function in motion processing. Preliminary experiments show that in flies expressing TeTxLC in all VS cells, compensatory head movements in response to vertical motion are unaffected (A. Keller and M. Heisenberg, in progress). Whether the function of the VS cells needs to be reconsidered or whether TeTxLC does not fully block their function remains to be seen.

B. Motor systems

1. Food intake

The most prominent muscles in adult *Drosophila* are used for locomotion (leg and wing muscles), but a variety of additional muscles are necessary for many processes such as altering the shape of the abdomen (hypodermal muscles of the abdomen), generating the hemolymph flow (heart muscles), and feeding (muscles of the digestive system). The process of feeding involves a series of motor functions (Dethier, 1976). The first response initiated by gustatory input is the extension of the proboscis, followed by spreading of labellar lobes and sucking of food. These sucking movements are likely to be generated by contractions of the three pharyngeal dilator muscles (Miller, 1950). Because the adult pharyngeal motor neurons mediating these contractions have been identified, the function of the different pharyngeal muscles in the process of feeding can be studied with appropriate P[GAL4] lines.

Tissot *et al.* (1998) used the P[GAL4] line MT26 to drive TeTxLC expression in three pairs of pharyngeal motor neurons (PMNs). The cell bodies of these six cells are located in the subesophageal ganglion, where they have both ipsi- and contralateral dendritic arborizations. The motor axons establish extensive arborizations on the pharyngeal dilator muscle 11. When TeTxLC was expressed in these motor neurons, food uptake was not reduced when flies were allowed to feed for 1 h. When flies were allowed to ingest food for only 12 s, food ingestion in flies with blocked PMNs was reduced to about 25% of control values. This clearly demonstrates the role of PMNs in feeding. However, flies with blocked PMNs are still able to feed to a limited extent. Since innervation of muscle 11 by unlabeled neurons is unlikely, considering the relevant anatomical studies (Rajashekhar and Singh, 1994), the other large dilator muscle (12) and/or the small median pharyngeal muscle (10) is probably also involved in generating the sucking movements for food intake.

2. Olfactory jump response

Sweeney *et al.* (1995) constructed flies containing the UAS-tetanus-toxin light-chain gene with the initial idea to use it to disrupt specific behavioral responses. In the first example of the use of the system, a well-defined reflex response, the olfactory-induced jump (McKenna *et al.*, 1989), was examined. Targeting the TeTxLC expression using a P[GAL4] line 129, a highly reduced jump frequency was found (less than 9% compared to control flies). For the jump, at least two components are required to be intact: the sensory component (which is specifically the olfactory circuitry) and the motor output component (the ability of the fly to jump), which requires several nerves and muscles. To assess which cells might be responsible for the jump response defect, the expression pattern of the line P[GAL4]129 was determined. Expression was detected in sensilla of the antennal segment, in maxillary palps, in the proboscis, in a few cell bodies of the brain and ventral nerve cord, and in a leg nerve in each hemisegment. However, at the stage of this study, the authors did not pinpoint which impaired cells are responsible for generating the decrement in jumping and whether these cells constitute part of the sensory component or the motor component of the reflex. To pursue this characterization, the use of other selective P[GAL4] lines should permit analysis of the specific neuronal circuitry underlying this behavior.

C. Central brain function

In insects, as well as in vertebrates, several approaches have been developed to explore and understand central brain function. The majority of the central brain structures, by their deep location in the brain, are not as directly or easily accessible as the sensory system. Therefore, lesion techniques are of limited interest since generally they generate secondary effects. Concomitantly, they often necessitate working with immobilized animals, which for obvious reasons greatly limits behavioral analysis. In neuroethology, to address the questions of what structure subserves a certain behavior and how it influences this behavior, one would ideally like to examine several independent methods that disturb the structure of interest. Indeed, since all of them putatively could generate side effects elsewhere in the organism, effects that will need to be distinguished from the intended impairment, we hope that in each independent method, the associated side effects might differ, permitting the inferrence of function from the overlap in data. Via the molecular genetic techniques outlined above, new noninvasive tools for investigating the neural substrate of behavior have been made available. However, as with other intervention techniques such as lesions or drugs, because of the pleiotropy of expression of the majority of the genes, genetic manipulations also generate uncontrolled and undesirable side effects, particularly those that may be generated during the developmental process. Therefore, if we consider results obtained with this new genetic method that are similar to those generated using other methods, mutual validation may allow us to put forward a convincing argument for assigning a specific role to a given structure. In the sensory system, behavioral impairments in some cases have been directly correlated with electrophysiological defects (Reddy *et al.*, 1997). However, since electrophysiological approaches have not yet been applied in the *Drosophila* adult central brain, the comparison of results from various methods of disturbance have become a suitable experimental approach used to define and delimit the role in behavior of various structures within the brain.

1. Mushroom bodies

a. Olfactory learning and memory

In insects, the mushroom bodies (MBs) are the adult central-brain structures that have been most studied, particularly at the level of development (Yang *et al.*, 1995; Tettamanti *et al.*, 1997; Ito *et al.*, 1997; Armstrong *et al.*, 1998; Lee *et al.*, 1999; Boquet *et al.*, 2000), but also for their function (reviewed: Heisenberg,

1989, 1994; Davis, 1993, 1996). In *Drosophila*, MBs are constituted by about 2500 intrinsic neurons (Kenyon cells). They receive multimodal sensory input, preferentially from the antennal lobe to the calyx, and send axon projections diagonally to the anterior brain, where they bifurcate to form complex subsets of medial (β, β', γ) and dorsal (α, α') lobes (Crittenden *et al.*, 1998; Ito *et al.*, 1997; Lee *et al.*, 1999; Yang *et al.*, 1995). Although they have been implicated in a variety of complex behavioral functions (Heisenberg, 1998), they are principally known and studied for their requirement in the olfactory learning and memory process.

Studies relating the MBs to learning and memory in insects have employed various approaches in various insects. Lesions have been commonly employed in the nongenetically tractable insects such as *Periplaneta* (Drescher, 1960), ants (Vowles, 1964), and honeybees (Menzel *et al.*, 1974; Erber *et al.*, 1980). In *Drosophila*, on the other hand, a number of approaches have been used to study MB function, encompassing genetic studies based on mutations that disturb MB structure (Heisenberg *et al.*, 1985), chemical ablations (De Belle and Heisenberg, 1994), and use of mutations disturbing relevant biochemical pathways (for review, see deZazzo and Tully, 1995; Davis, 1993, 1996).

Interestingly, several genes involved in the cAMP pathway are preferentially expressed in Kenyon cells, in particular *dunce*, which encodes a cAMP phosphodiesterase (Nighorn *et al.*, 1991), *rutabaga*, which encodes a Ca^{2+}/calmodulin-dependent adenylate cyclase (Han *et al.*, 1992), and the cAMP-dependent protein kinase (PKA) (Drain *et al.*, 1991; Skoulakis *et al.*, 1993). More recently, the availability of the new binary enhancer-trap P[GAL4] system to drive effector genes has also been used to investigate MB function.

Based on their effects on the cell physiology, effector genes useful for studying MB function can be subdivided in four categories. First, it is possible to remove or ablate MBs by killing them with the expression of an apoptosis-inducing transgene, such as *reaper* or *hid* (McCall and Steller, 1997). As yet, no study that has used this approach to ablate the MBs in whole or part has been reported. A second category of effector genes can be expressed in order to disturb the physiology of the cell, by expressing a constitutive mutant form of a gene (or dominant gain-of-function allele) in order to disturb a well-characterized molecular pathway. In this way, Connolly *et al.* (1996) have shown that the MB-directed expression of the constitutively active form of the G-protein α subunit, ectopic expression of which is known to disrupt the level of the cAMP produced in a cell (Quan *et al.*, 1991), disturbed the process of olfactory learning. A third category of effector genes, recently developed, is based on the targeted expression of a wild-type copy of a gene in an attempt to rescue function in a limited set of cells, in a background that is otherwise mutant for this gene. Indeed, the widespread expression of several genes has always limited the interpretation of the learning and memory defects generated by their mutation. In this inverted approach, which relies on the correct function of defined brain regions, Zars *et al.*

(2000a), using a set of P[GAL4] enhancer-trap lines, have been able to rescue the *rutabaga* mutant memory defect by the tissue-specific expression of a type 1 adenylyl cyclase (*rutabaga* gene). Comparing expression patterns of rescuing and nonrescuing lines suggests that the restitution of the *rut* product in the γ lobes is sufficient to rescue the *rutabaga* learning and memory phenotype. These results represent the first attempt to assign a specific function to a subset of Kenyon cells and also represent a functional subdivision of the MBs. The fourth category of effector genes refers to the targeted expression of TeTxLC, the action of which blocks synaptic transmission, leading to silencing of the neurons. To date, the expression of TeTxLC in the MBs has not been used to study olfactory memory. However, it should be noted that until now, with the exception of the restitution of the *rut* gene in the γ lobes (Zars *et al.*, 2000a), no studies of MB function have led to a precise assignment of a role for the various subsets of Kenyon cells. Likewise, we are still none the wiser about how the Kenyon cells code mnesic odor identity. Moreover, it should be stressed that in *Drosophila* the output pathway of the MBs is still only partly characterized (Ito *et al.*, 1998); therefore, the mechanism by which the MBs as a learning and memory center influence these downstream premotor pathways remains vague.

Many enhancer-trap P[GAL4] lines have been reported to have an expression in the MBs (Yang *et al.*, 1995; Boquet *et al.*, 2000). However, to date, several of these lines have been unusable in studying the functional dissection of various subsets of Kenyons cells because in combination with tetanus toxin, the flies die before they reach an adult stage. It is hoped that the development of toxigenetics, especially in a second generation (see details later), which would permit the temporal control of the expression of the effector gene, will allow the use of these lines to unravel the function of the various subsets of Kenyons cells, as well as go some way toward mapping the link between MBs and other parts of downstream brain circuitry.

b. Context generalization

Another aspect of the requirement of the MBs for normal behavior has been revealed using a complex piece of apparatus referred to as the flight simulator. Briefly, a tethered fly is glued by its head and thorax to a small copper wire hook and suspended from a torque meter. The suspended fly is surrounded by an arena that can be rotated by an electric motor. The arena is illuminated from behind the fly and carries four black patterns, two upright and two inverted T-shapes, in an alternating sequence on its wall (a T sits at each of the four points of the compass from the vantage point of the fly). The angular velocity of the arena is proportional to but directed against the fly's yaw torque, generated as the animal flies. Thus, instead of rotating the fly in a stationary surrounding, the fly's yaw torque rotates the panorama around the stationary fly (thus simulating flight). To test visual learning, as a negative reinforcement the fly is heated by an infrared-light beam when it attempts to fly toward a particular pattern in the arena. When the fly corrects its course to a permissive pattern, the infrared beam is intercepted by a computer-driven electric shutter (for a more detailed description of the apparatus and a schematic drawing, see Wolf *et al.*, 1998; Li *et al.*, 1999). Although it has been shown that MBs are dispensable for visual, tactile, and motor learning (Wolf *et al.*, 1998), they have been shown to be required for a more subtle visual learning task, known as "context generalization" (Li *et al.*, 1999). In the visual-learning paradigm just described, a tethered fly can be conditioned by heat to prefer certain flight orientations and avoid others (Wolf and Heisenberg, 1991). With this operant learning paradigm, the authors systematically changed certain features of the experimental setup (context) between training and test. The manipulated context was the illumination of the panorama. Then, testing the influence of the color changes of the panorama on memory retrieval, they showed that MB-disturbed flies do not exhibit memory, as do control wild-type flies. Here also, the authors used three independent and non-invasive approaches to impair MB function. In addition to HU ablation and the use of the mbm^1-mutation, targeted expression of TeTxLC in a subset of intrinsic neurons of the MBs (here, restricted to the α/β type, by the P[GAL4]17D) led to a block in memory retrieval. Again, targeted expression of the toxin has been revealed to be specific enough to be used as tool to dissect complex behavior and may represent the first example of a functional dissection of the internal circuitry of MBs.

c. Locomotor activity

In contrast to the well-documented implication of the MBs in olfactory learning, much less is known about MB function in motor control. Various approaches in different organisms have suggested a role of the MBs in the control of the motor program or in the level of motor activity. For example, earlier studies have shown that unilateral MB ablation in *Cecropia* larvae leads to aberrant cocoon spinning (van der Kloot and Williams, 1953), whereas in adult male crickets MB lesions lead to an elevated general motor activity (Huber, 1955). Similarly, locusts and bees with MB lesions show a general increase of behavioral activity (for review, see Erber *et al.*, 1987). In *Drosophila*, genetic lesions that give rise to central brain structural defects, such as *mushroom body miniature¹* (*mbm¹*) mutants, also exhibit increased locomotor activity (reported in Heisenberg *et al.*, 1985). In addition, flies with chemically ablated MBs have been found to be more active than control flies, in conjunction with investigations of a role for MBs in maintenance of circadian rhythm (De Belle *et al.*, 1996). Taken together, these results from various sources strongly suggest that MBs may play a role in the generation, regulation, or coordination of motor programs. Martin *et al.* (1998, 1999a) have developed a new quantitative paradigm in which the locomotor activity of a single fly is automatically recorded over a time period up to 7 h.

Several parameters of locomotor activity are extracted to reveal temporal patterns, including the total amount of activity and its time course, as well as the duration of the successive episodes of activity/inactivity. In this paradigm, flies with disturbed MBs exhibit a higher level of locomotor activity (compared with control flies) and do so by increasing the duration of their episodes of activity to the detriment of the duration of their inactivity episodes (Martin *et al.*, 1998). These results confirm that the MBs do indeed play a role in the regulation of the level of general locomotor activity. More importantly, the authors have compared three different and independent approaches to disturbing function in the MBs, namely, chemical ablation [using hydroxy urea (HU)], genetic lesions (using mutations affecting Kenyon cell differentiation), and the targeted expression of TeTxLC, and have shown that the three methods of disruption lead to a similar phenotype: an increase in locomotor activity. These results indicate that targeted expression of TeTxLC is sufficiently specific to be used to dissect central brain function. The availability of a large number of enhancer-trap P[GAL4] lines with various expression patterns within the MBs (for example, Yang *et al.*, 1995; Boquet *et al.*, 2000) is surely one of the most promising approaches to dissecting the function of the different sets of Kenyon cells within the MBs, either for their implication in learning and memory or for their role (though it may be indirect) in the control of general locomotion.

2. Central complex

While the MBs have been intensively investigated, probably because they receive connection directly from known sensory modalities, less attention has been paid to the central complex (CC), another complex central brain structure principally composed of a regular array of repetitive fibers. The CC is an unpaired structure located deep in the middle of the brain. It is subdivided in four substructures: protocerebral bridge (pb), fan-shaped body (fsb), ellipsoid body (eb), and noduli (no) (Hanesch *et al.*, 1989). In contrast to the MBs, the CC is not solely composed of intrinsic neurons derived from a common set of a few neuroblasts (precisely, only four for the MBs; Ito *et al.*, 1997), but rather is a neuropil that represents the synaptic connections between several heterogeneous groups of neurons. Those neurons have their cell bodies spread in various parts of the protocerebrum. Although many of them have been described (Hanesch *et al.*, 1989; Martin *et al.*, 1999b, in press; Renn *et al.*, 1999b) several remain to be characterized. The heterogeneity of the cytoarchitectural structure of the CC leads to difficulty in dissecting both the development and structure of this intriguing structure. Therefore, toxigenetics applied to brain-behavioral studies of the CC have special utility in dissecting the function of this structure at the neuronal level, by specifically targeting subgroups of neurons.

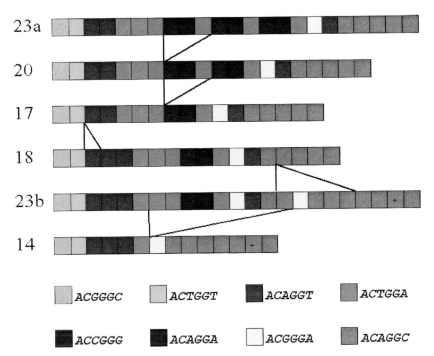

Figure 4.1. Evolution of the Thr-Gly length alleles of the *per* gene in *D. melanogaster*. Each box represents a pair of codons specifying a Thr-Gly repeat. The different colors represent the different repeat sequences due to variation at third-base positions within codons. The number of Thr-Gly repeats found in each variant is shown on the left. Note the existence of two different alleles coding for 23 Thr-Gly pairs. The various alleles can be derived one from another by a series of insertion/deletion events as shown in the figure. The plus sign represents a nucleotide substitution turning the ACTGGA repeat into ACTGGC (Costa *et al.*, 1991).

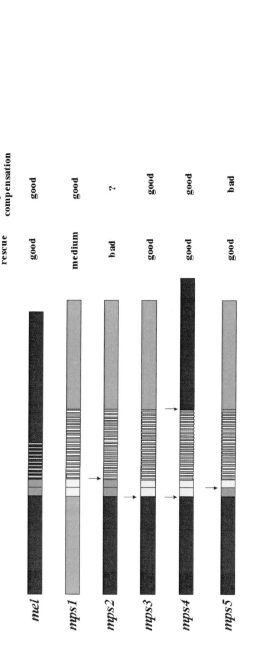

Figure 4.3. Interspecific chimeric constructs. This figure illustrates the behavioral results obtained after the transformation of per^{01} arrhythmic mutants with different chimeric constructs containing fragments of the *per* genes of *D. melanogaster* and *D. pseudoobscura*. per^{01} transgenic flies carrying a construct of the *D. melanogaster* gene show good rescue of its circadian activity rhythms and a good temperature compensation for cycle duration (ca. 24 hr). The rescue obtained with the *mps1* construct containing regulatory sequences of *D. melanogaster* and the whole coding sequence of *D. pseudoobscura* present medium levels of rescue (Petersen *et al.*, 1988; Peixoto *et al.*, 1998), probably due to differences in the N-terminal sequences that contain the PAS dimerization domain (Piccin *et al.*, 2000). Constructs *mps2* and *mps3* are very similar. Both contain the first half of the *D. melanogaster* coding sequence and the second half of the *D. pseudoobscura* sequence (including the repetitive domain). However, they differ in the exact position of the chimeric junction. In *mps3* the *D. pseudoobscura* repeat region is flanked by its own coevolved upstream sequences, and this construct worked very well, giving a high level of rescue of rhythmicity and good temperature compensation. On the other hand, in *mps2*, which gave very poor rescue, the *D. pseudoobscura* repeat region is flanked by the *D. melanogaster* upstream sequence. The *mps4* construct, which has the *D. pseudoobscura* repeats and its upstream flanking region embedded in a *D. melanogaster* gene, shows that the coevolution between the repeat region and its flanking sequences does not include the C-terminal end of the protein. *mps5* shows that dividing the coevolved flanking sequence causes disruption to the temperature compensation mechanism of the circadian clock.

a. Maintenance of locomotor activity

Early attempts to analyze central-brain function in grasshoppers and crickets, using small lesions and focal electrical stimulation, have assigned a behavior-initiating or maintaining function to the CC (Huber, 1955, 1960, 1963, 1965; Otto, 1971; Wahdepuhl and Huber, 1979; Wahdepuhl, 1983; reviewed in Homberg, 1987). In *Drosophila*, studies using structural-brain mutants have suggested that, among probably several putative functions, the first emerging function for the central complex is that it is a higher control center for locomotion (Strauss and Heisenberg, 1993). Mutants with defects in the architecture of the CC are impaired in a variety of walking parameters: walking speed, straightness of walking, time course of walking activity, leg coordination during turns, and start–stop (Leng and Strauss, 1996, 1997; Wannek and Strauss, 1997). Surprisingly, in one of the mutant strains, *no-bridge* (*nob*), the flies from time to time fell into a "comatose-like state" during which they could not walk but still maintained a standing posture and performed occasional erratic leg movements. Using genetic mosaics to pinpoint the precise locus of the genetic lesion, the neuronal defects found in these mutants were found to reside in the head and are correlated with the structural defects in the CC (Strauss *et al.*, 1992; Strauss and Heisenberg, 1993). Additionally, these same CC mutants present some defects in olfactory and visual learning tasks (Heisenberg *et al.*, 1985) and in various properties of visual flight control (Ilius *et al.*, 1994). Finally, activity labeling of neurons with radioactive deoxyglucose has also supported a role for the CC in locomotor control (Bausenwein *et al.*, 1994). However, because these mutants have not yet been molecularly characterized, and thus the precise identity of affected neurons and the nature of the molecular lesion are still unknown, the resolution of the defect has been assigned solely at a gross structural level.

Continuing these investigations, taking advantage of the enhancer-trap P[GAL4] system to drive the expression of TeTxLC, Martin *et al.* (1999b) initiated dissection at the neuronal level of the internal circuitry of the central complex, in relation to locomotor activity, comparing results obtained from structural mutants missing parts of the CC with those obtained from strains in which TeTxLC blocked various populations of CC neurons. Two independent small groups of neurons were targeted by TeTxLC expression: first, a novel cluster of large-field tangential neurons of the fan-shaped body, in which only five to six neurons arborize in a layer of the upper part of the fan-shaped body; and second, a group of four to eight neurons, the pb-eb-no (Hanesch *et al.*, 1989), linking the protocerebral bridge, the inner part of the ellipsoid body, and the ipsilateral noduli. In targeting each of these groups with TeTxLC expression, a common parameter of walking activity was altered. Although the number of episodes of activity was normal, their duration was critically reduced, suggesting that the maintenance of locomotor activity is affected and not its initiation. As suggested by the anatomical data, these two clusters of neurons probably belong to a common circuitry, possibly forming neuronal networks locally.

In summary, disturbing the posterior region of the CC leads to a phenotype that contrasts with the phenotype observed when the MBs are disturbed in that disruption of posterior CC function reduces locomotor activity. These results, obtained using noninvasive methods on freely walking animals, parallel findings in orthopterans 40 years ago, using intervention methods on stick insects (Huber, 1960). Moreover, it is interesting that the toxigenetic technique has permitted the inhibition of a very small number of neurons in the brain, as few as 5 or 6, and this is enough to cause a phenotypic impairment. These results represent the first study of dissecting CC substructures at a cellular level. Unfortunately, we do not have yet at our disposal P[GAL4] enhancer-trap lines that label and block other precisely described neuronal fibers of the CC, such as the horizontal fiber systems (HFSs) and vertical fiber system (VFS), as well as both types of pontine neurons (neurons connecting adjacent segments and adjacent layers) (Hanesch *et al.*, 1989). Yet, one can hope that such P[GAL4] lines will become available in the near future. If so, this should permit the exploration in more detail of the respective roles of each group of neurons of the CC substructures as well as interactions between these various groups. It is hoped further that a picture may emerge of how locomotor activity is regulated neurophysiologically, and in a wider context how the motivational state for walking is regulated. Finally, this paradigm of locomotor activity in which the walking versus nonwalking state is quantified could also possibly be considered a new "decision making" experimental system, since, whatever the fundamental reason that a fly moves at a given time, it certainly reflects the fly's decision to walk or not walk.

b. Fractal structures in temporal patterns of locomotor activity

As suggested by anatomical data and by the similarity of observed phenotypes, TeTxLC-blocked CC neurons probably form a local network linking the protocerebral bridge, the fan-shaped body, and the noduli. Based on its cytoarchitectural organization, the CC can also be subdivided into a posterior part, comprising the pb and the fsb, and an anterior part, principally the ellipsoid body. Enhancer-trap P[GAL4] lines with specific expression in the ellipsoid body have also been used to express the tetanus toxin in the search for a role for this structure (Martin *et al.*, in press). Although the main neurons forming the ellipsoid body are the ring neurons (Hanesch *et al.*, 1989), as in cases involving other parts of the CC, use of P[GAL4] enhancer-trap techniques have revealed additional groups of neurons with projections into the ellipsoid body (Renn *et al.*, 1999b; Martin *et al.*, in press).

Curiously, blocking neurons that arborize in the ellipsoid body leads to an intriguing phenotype: the temporal pattern of locomotor activity does not present a power-law distribution of its intervals (a property often referred to as fractal), such as that seen in wild-type flies (Martin *et al.*, 1999a,b, in press). Power-law distributions have been discovered in the time course of a variety of physiological processes, such as electroencephalographic time series (Accardo *et al.*, 1997; Lutzenberger *et al.*, 1995; Martinerie *et al.*, 1998; Le Van Quyen *et al.*, 1999), heartbeat (Goldberger and West, 1987; Elbert *et al.*, 1994), and the temporal dynamics of the clinical manifestation of certain psychiatric disorders, such as physical tics in Gilles de la Tourette syndrome (Peterson and Leckman, 1998). Even in *Drosophila*, fractal temporal variability has been found in timing of behaviors such as feeding (Shimada *et al.*, 1993) and movement (Cole, 1995). However, in all those systems, no neuronal bases or neuronal networks underlying such highly structured behavior have been described. In *Drosophila* we have the tools to disturb this power law. Since the disruption of the power law is only observed for lines driving expression in neurons that have projections into the ellipsoid body, a hypothesis can be put forward that the fine tuning of the organized temporal pattern of locomotor activity occurs in the ellipsoid body. It follows that the intricate cytoarchitecture of the ellipsoid body, with its ringlike arborization, may be the seat of such behavioral fine tuning.

3. Circadian rhythms

Several other behavioral sequences have also been intensively investigated in relation with brain structure. Since the pioneering work of Konopka and Benzer (1971), the search for the neural bases of biological rhythms has received much attention, and this work has been reviewed many times (e.g., Rosbash and Hall, 1989; Kyriacou and Hall, 1994; Hall, 1990, 1995; Dunlap, 1999). Nonetheless, the main molecular players in this pathway, genes such as *period* (*per*) and *timeless* (*tim*), were cloned and characterized before the neuronal brain circuitry implicated in the circadian rhythms was known. Indeed, only relatively recently, and mainly based on the use of detailed immunohistochemistry, have the main brain neurons constituting the putative circadian pacemaker neurons been well described and their projections in the brain precisely traced (Helfrich-Förster and Homberg, 1993; Helfrich-Förster, 1995, 1997, 1998; Kaneko *et al.*, 1997; Kaneko, 1998; Kaneko and Hall, 2000). Six clusters of brain neurons expressing the two major circadian genes, *per* and *tim*, have been described. Three of these clusters are located in the dorsal protocerebrum (DN: dorsal neurons), while the other three are located in the anterior lateral cortex. Those lateral neurons (LN) are also divisible into dorsal (LN_d) and ventral (LN_v) clusters. Moreover, the LN_v coexpress a neuropeptide referred to as pigment-dispersing factor (PDF) (Helfrich-Förster, 1995; reviewed by Helfrich-Förster *et al.*, 1998). Using a detailed anatomical and behavioral study of individual *disconnected* mutants, and a detailed examination of their PDF immunoreactivity, Helfrich-Förster (1998) has shown that the robust circadian rhythm of *Drosophila* requires the presence of lateral neurons. These findings, augmenting earlier analogous ones (Ewer *et al.*, 1992; Frisch *et al.*, 1994), suggest that the LN and especially the LN_v appear to be critically involved in the circadian control of locomotor activity (reviewed by Kaneko, 1998). Therefore, the LN and LN_v are by far the best candidates for central brain circadian pacemaker neurons in *Drosophila*. However, the biological significance of the DN and LN_d remains largely unknown.

To confirm the role of LN_v as principal circadian pacemaker, and concomitantly to investigate and discern the role of other potentially relevant circuitry (such as the DN and LN_d neurons) underpinning circadian rhythms, the targeted expression of the apoptosis-inducing gene *reaper* as well as TeTxLC was employed. Two different groups of workers, using similar strategies, have arrived at similar conclusions. Chronologically, Renn *et al.* (1999a) have used a *pdf*-GAL4 driver (the transcription factor GAL4 placed downstream of the *pdf* regulatory element gene, designed by Park and Hall, 1998; Park *et al.*, 2000) to express the cell-death genes UAS-*rpr* or UAS-*hid*. Flies with selectively ablated *pdf*-containing (LN_v) neurons are arrhythmic, supporting the assignment of LN_v neurons as the principal circadian pacemaker. However (and surprisingly), expression of TeTxLC strictly restricted to the LN_v neurons, using the same *pdf*-GAL4 driver, has only a mild effect on behavioral rhythmicity (Kaneko *et al.*, 2000). Additionally, removing the same LN_v neurons, by using a novel enhancer-trap P[GAL4] line (P[GAL4]1118) to drive a newly described cell-death gene, UAS-Bax2 (Gaumer *et al.*, 2000), yields arrhythmic flies, whereas in parallel, blocking the same LN_v neurons with the tetanus toxin has only a mild effect (Blanchardon *et al.*, 2001). One of the most plausible explanations for this clear-cut difference between removing a group of neurons and blocking them with TeTxLC may be that since the toxin is known to block "fast" chemical synaptic transmission, and PDF is a neurohormone likely to be secreted by a related but different mechanism (discussed in Section II.D), PDF secretion is insensitive to TeTxLC expression. This suggests that the release of the PDF neuropeptide may be mediated by an alternative mechanism, one that does not involve or is insensitive to the loss of n-synaptobrevin, the only currently identified substrate of TeTxLC in *Drosophila*.

D. Miscellaneous

1. Using *Drosophila* to study the effects of substances of abuse

The biological components underlying addictive processes in humans, such as sensitization to psychostimulant drugs of abuse, is another aspect of behavior that has been studied in *Drosophila*, with the eventual aim of taking advantage of the powerful forward genetics afforded by this model system. Hirsh and co-workers (McClung and Hirsh, 1998, 1999) have shown that *Drosophila* exhibit stereotypic motor responses to cocaine exposure similar to those seen in vertebrates, offering the possibility that Drosophila could be used as a new and tractable model to study cocaine sensitization. In humans (Kalivas et al., 1998) and in vertebrate animal models (Post and Rose, 1976; Kalivas et al., 1988), this addictive process has been thought to be partially mediated by dopaminergic regions of the brain. Additionally, the sensitization observed when mammals are exposed to cocaine implicates presynaptic changes within the dopamine neurons themselves. In vertebrates, aminergic neurons and amine synthesis and release are negatively regulated by $G\alpha_i$-coupled autoreceptors (Pothos et al., 1998). Interestingly, in Drosophila, the modulation of cell signaling in dopamine and serotonin neurons, by P[GAL4]-mediated targeted expression of either a stimulatory ($G\alpha_s$) or inhibitory ($G\alpha_i$) $G\alpha$ subunit, by the use of the GAL4 expression under the Ddc (dopa decarboxylase) promoter (pDdc-GAL4), blocked the behavioral sensitization to repeated cocaine exposures (Li et al., 2000). Additionally, blocking the same neurons with TeTxLC expression leads to a similar phenotype (Li et al., 2000). This suggests that, in a process that strongly resembles the situation in mammals, Drosophila sensitization requires the modulation of transmitter release presynaptically within monoamine-containing neurons (Li et al., 2000). This experimental paradigm also revealed the similar effects of disturbing the same specific neurons with different methods, strengthening the conclusions that these dopaminergic neurons are a focal point for sensitization.

In another behavioral paradigm, Heberlein and co-workers have studied the effects of acute and chronic ethanol exposure on Drosophila in an effort to dissect the molecular mechanisms underlying the various responses (which in flies are very similar to those seen in mammals) to this substance (Bainton et al., 2000; Moore et al., 1998; Scholz et al., 2000). Drosophila develop ethanol tolerance after a single exposure, as evidenced by their increased resistance to sedation and delayed slowing of locomotor behavior upon a repeated exposure to ethanol vapor (Scholz et al., 2000). Use of central brain structural mutants and targeted expression of TeTxLC demonstrate that structural integrity of the central brain is necessary for the development of ethanol tolerance, notably specific parts of the central complex and possibly the mushroom bodies.

There is strong evidence that ethanol tolerance in mammals is a learned response and that particular brain structures play a role in this process (Le et al., 1981). In flies, the involvement of molecules previously implicated in learning processes in this response of flies to ethanol (Moore et al., 1998) strongly suggests that the effects wrought by ethanol exposure on flies and mammals are brought about by similar mechanisms. The astonishing similarities at many levels, behavioral and mechanistic, between Drosophila and vertebrates exposed to various substances of abuse raise the exciting possibility that Drosophila can be used to identify novel molecular targets for sensitization and tolerance to these drugs.

2. Courtship behavior

Courtship behavior plays a key role in animal life by allowing both sexes to find a mate and thereafter proceed into a behavioral ritual that ends in successful copulation. Although courtship has evolved as a "complex behavior," the ritual in flies is composed of a string of "fixed" behavioral sequences, which can be broken down into observable sterotypical steps. These steps have been previously described (reviewed in Hall, 1994a) and normally occur in the following order: (1) orientating of the male to the female, (2) tapping of the female by the male with his prothoracic legs, (3) tracking the female when she moves, (4) singing, associated by the vibration of one or the other of the male's wings, (5) male licking of the female genitalia, (6) attempted copulation, and finally (7) copulation.

The nature of this behavioral sequence makes it an attractive model system with which to investigate a neural basis for such "hardwired" behavior. Indeed, a huge number of studies have been performed in order to identify behaviorly relevant genes and dissect the underlying circuitry (see reviews: Taylor *et al.*, 1994; Hall, 1994a; Yamamoto *et al.*, 1997). However, as with other reported behavioral studies, such as circadian rhythms, it would probably not be an understatement to say that the recent analysis of genes involved in courtship has preceded the precise characterization of the neuronal circuitry underlying the various aspects of courtship behavior. Many studies of fly courtship have led to the identification of a panoply of genes implicated in a variety of molecular pathways, but in the final analysis, very few have been revealed to be courtship specific (reviewed in Hall, 1994a).

Nevertheless, studies employing involved genetic techniques have been used to map brain-specific foci for sex-specific behavior. The gynandromorph technique (which allows the generation of somatically sexually mosaic flies, in which some parts of the fly are male and other parts female) has been utilized to locate anatomical sites that are required to be one sex or the other for the correct performance of a behavioral sequence (Hotta and Benzer, 1972). In this manner, the first maps of sex-specific foci for sequential action patterns of courtship have been traced (against a background of previous establishment of the requisite mosaic methods: Hotta and Benzer, 1976; Kankel and Hall, 1976). Of particular interest, in the male adult brain, neurons of the dorsal posterior part of the brain (possibly corresponding to the posterior part of the pars intercerebralis) have been shown to be necessary to be male to allow a male to initiate courtship with a female (Hall, 1977, 1979). However, the low frequency of gynandromorphy and the inability of the technique to control the precise location of the mosaic have limited its power.

As an alternative to the gynandromorph technique, the use of the targeted expression of the P[GAL4] system has permitted a more precise and repeatable feminization of specific brain structures, using the targeted expression of a feminizing gene, *transformer* (*tra*) (Ferveur *et al.*, 1995; O'Dell *et al.*, 1995). In these studies, parts of the antennal lobes as well as parts of the MBs have been shown to be implicated in courtship behavior because they are required to be male for correct partner recognition/sexual discrimination. In a continuance of this study, using several enhancer-trap P[GAL4] strains to drive the *tra* gene, Ferveur and Greenspan (1998) further refined the map of courtship foci in the brain. With the same technique, a few neurons located in the anterior part of the pars intercerebralis have been identified as being responsible for the sexual dimorphism of locomotor activity in *Drosophila* (Gatti *et al.*, 2000). It should now be possible, with the help of different enhancer-trap P[GAL4] lines driving precise and restricted expression patterns, to target the expression of TeTxLC (or the feminizing *tra* gene) in order to block specific neurons of the posterior part of the pars intercerebralis. Such a line of study might refine the neural cartography of the pars intercerebralis.

Although courtship behavior is highly stereotyped, some of its aspects have also been shown to be experience-dependent. Studies with memory mutants support this interpretation (Siegel and Hall, 1979; Siegel *et al.*, 1984; Kyriacou and Hall, 1984). More particularly, in females, experience-dependent modification is concerned with receptivity, whereas in males, it relates particularly to sexual orientation (or partner discrimination). For example, young males just after eclosion sexually attract mature males, inducing intense homosexual courtship (e.g., McRobert and Tompkins, 1983; Vaias *et al.*, 1993). The experience of the mature with the immature males results in a distinct alteration in courtship response, which is thus considered an experience-dependent courtship modification (Gailey *et al.*, 1982). Manipulations of the MBs, by HU ablation, has resulted in the inability of mature males to cease courtship when placed with immature males (Neckameyer, 1998), suggesting that the MBs are important in the neuronal circuitry sustaining this simple form of learning or habituation. A second aspect of male experience-dependent modification concerns courtship conditioning, in which negative courtship-related experience associated with mated females causes males to subsequently reduce their courtship toward virgins. In normal males, this negative courtship experience establishes a long-term memory lasting for up to 9 days, whereas in mushroom-body HU-ablated males, courtship conditioned memory dissipates within 1 day (McBride *et al.*, 1999).

As yet, no studies that employ TeTxLC expression to block various parts of the circuitry implicated in the different components of courtship behavior have been reported. However, Heimbeck *et al.* (2000) have used the P[GAL4] line GH146 to target the expression of TeTxLC specifically to the antennoglomerular tract. While the response to certain odors is reduced in a T-maze assay, compared to control lines, and the gustatory responses are normal in proboscis extension reflex assay, TeTxLC expression almost completely abolishes the male's courtship. This suggests that the projection neurons (PNs) joining the antennal lobes to the MBs are essential for the conveyance of information (here, most likely the detection of female-specific volatile pheromones, a highly specialized subset of olfactory stimuli) to the higher integrative brain centers, in particular the MBs, reinforcing the implication of MBs in this process. Yet no direct blockage of the different subsets of Kenyons cells of MBs, by TeTxLC expression, has been performed and examined for its consequence upon courtship. Additionally, whether the simple form of learning seen in mature males after courtship of young males—as well as short- and long-term memory in males—can be impaired by the expression of TeTxLC in MBs, in P[GAL4] lines 17D, 201Y, or H24, for example (cf. Martin *et al.*, 1998), remains to be studied.

3. Tetanus toxin in other species

In other animals, especially rat, tetanus toxin has been used as a tool to impair neurophysiological function and notably to induce epilepsy. In the "rat tetanus-toxin model of epilepsy" (Finnerty and Jefferys, 2000), the whole molecule of the tetanus toxin, both the heavy and light chains, is directly injected into the brain, focusing on the hippocampal region. This leads to a chronic defect of intermittent epileptic seizures, which is preceded by a particular 9- to 16-Hz oscillation of field potentials that are normally synchronized between the right and left dorsal regions of the hippocampus. An alternative protocol has also revealed that rats which have recovered from an experimental limbic epilepsy present a reduction in the exploratory response to novelty (Mellanby *et al.*, 1999). If such an "epilepsy model" similarly occurs in mouse, it might then be possible to achieve both regional and temporal control of expression of the catalytic light chain alone, not unlike the system in use in *Drosophila*, in this case employing various brain (hippocampal)-specific promoters (e.g., Mayford *et al.*, 1996; Tsien *et al.*, 1996). Using such a system, the foci for both the 9- to 16-Hz oscillations and a reduction in the exploratory response to a novel object may be defined and the role of the hippocampus in epileptic seizures explored.

V. CONCLUSIONS AND PERSPECTIVES

We have reviewed the action and mechanism of TeTxLC, the methodology employed for its specific expression—specifically the binary enhancer-trap P[GAL4] system—as a new tool for studying the neurobiology of behavior. We have also reported the use of this system to dissect diverse components of physiological systems and/or behavior. This present review demonstrates the versatility and utility of the TeTxLC targeted-expression system. The reliability of the technique is supported strongly by the demonstration, via electrophysiology, of the full block of evoked synaptic transmission (Sweeney *et al.*, 1995), which silences the neuron in which TeTxLC is expressed. Moreover, the electrophysiological evidence of blocked synaptic transmission caused by TeTxLC expression and the direct link to the disruption of an associated physiological process, the resistance reflex (Reddy *et al.*, 1997), also add to the credibility of the system. Additional strength of the system is afforded by the ready visualization of expressed TeTxLC by direct immunocytochemistry. Indeed, the expression of the toxin in fly brains can be precisely detected following behavioral analysis. In related behavioral studies, the expression pattern of the effector gene is generally deduced indirectly from the use of a coexpressed reporter gene, such as β-galactosidase, green fluorescent protein (GFP), or tau [e.g., expression of *transformer* gene (Ferveur *et al.*, 1995; Gatti *et al.*, 2000; or *rutabaga* adenylyl cyclase gene (Zars *et al.*, 2000a,b)]. Near certainty can thus be achieved in assigning a role to a neuron or group of neurons, and discrepancies avoided caused by differences between the expression of the effector gene and the reporter gene used to reveal an expression pattern. However, this does not lessen a second problem that, very commonly, several enhancer-trap P[GAL4] lines exhibit, in that they often have expression elsewhere than one desires (also, spatial and temporal expression of the P[GAL4] may vary during development; see later discussion, leading to possible secondary effects that need to be distinguished from the effects of the intended impairment.

The versatility of the system has been demonstrated by the many uses it has been put to in the study of fly behavior. Many studies have now employed TeTxLC expression. First, the use of TeTxLC has proved very effective in studying relatively simple behaviors, ranging from the reflex responses—such as the resistance reflex (Reddy *et al.*, 1997) and the olfactory jump response (Sweeney *et al.*, 1995)—to the role of muscle contraction in feeding (Tissot *et al.*, 1998). Similar results have been found using toxin expression as a tool for studying more complex behaviors, such as the functioning of olfaction and gustation in larvae (Heimbeck *et al.*, 1999), olfaction in adult flies (Devaud and Ferrus, in preparation), mechanoreception and proprioception (Keller *et al.*, in press), central brain function, [e.g., context generalization (Li *et al.*, 1999), control of locomotion (Martin *et al.*, 1999a,b)], and circadian rhythms (Kaneko *et al.*, 2000; Blanchardon et al., 2001). Moreover, TeTxLC expression finds a use in dissecting central brain reaction to substance abuse (Li *et al.*, 2000; Scholz *et al.*, 2000). Toxigenetics has thus significantly contributed to bringing new insights into the dissection of the neural bases of behavior.

However, as with all tools employed in the laboratory, there is always room for improvement. It has been demonstrated that the observed effects of TeTxLC expression arise from silencing of the expressing neuron and are most likely not due to secondary developmental defects generated during the developmental process [though in three cases some slight developmental defects have been described at the synapse (Baines *et al.*, 1999; Heimbeck *et al.*, 1999; Hiesinger *et al.*, 1999)]. Many enhancer-trap P[GAL4] lines may drive expression in nonrelevant tissues during development, though the adult expression pattern may appear precise in the brain structure of interest. This makes such P[GAL4] lines unusable, since in combination with TeTxLC, death at a stage earlier than desired may be the outcome. To circumvent this problem, the establishment of a conditional and temporarily inducible system for TeTxLC expression should permit the use of such P[GAL4] lines for behavioral studies. Some attempts to achieve a combination of temporal and spatial control have been developed and are being refined or extended. So far, two versions of an inducible temporal and targeted expression system based on tetracycline have been developed (Bello *et al.*, 1998; Bieschke *et al.*, 1998). However, the integration of the use of the tetanus-toxin gene with this tetracycline-controlled system has not been assessed, and much work must be carried out before the combination of these two systems will be usable for behavioral studies.

Another strategy for achieving temporal control utilizes the yeast FLP/FRT recombination system. Between a GAL4 recognition sequence (UAS) and a TeTxLC cDNA, a cassette is placed that contains a heterologous gene (*miniwhite*), flanked by two FLPase recombination enzyme recognition sequences. Briefly, after induction of an FLPase enzyme by heat shock, the heterologous sequence is excised by recombination, leaving just one FRT site, between the UAS sequence and the TeTxLC cDNA. In the presence of GAL4, TeTxLC is then expressed (Keller *et al.*, 2000; Sweeney *et al.*, 2000; Smith *et al.*, 1996). Although the FLP/FRT recombination system is traditionally used in mitotically dividing cells to swap chromosome arms (Theodosiou and Xu, 1998), the placing of the FRT sites in *cis* may allow recombination in terminally differentiated cells. In its current usage, flies containing such a construct are heat-shocked at a late stage of development to turn on TeTxLC expression in a defined pattern, thus circumventing lethality normally caused when a standard UAS-TeTxLC construct is used with the same P[GAL4] line (Keller *et al.*, 2000). This second generation of the toxigenetic tools should allow the use of a much larger number of enhancer-trap P[GAL4] lines, as well as temporal control over neuronal activity in specific brain structures in behavioral studies.

The directed expression of TeTxLC, under the control of the enhancer-trap P[GAL4] system, disturbs the physiology of the neuron by blocking synaptic transmission leading to the silencing of the TeTxLC-expressing neurons; which neurotransmitter the neurons employ need not be known beforehand. This tool thus seems to be more amenable to the investigation of how a behavior is neuronally integrated in terms of circuitry, neural assembly, and/or neuronal networks. This modus operandi stands in opposition to the concept of "clone and sequence" and its corollary, the "behavioral" gene, both of which serve to determine, at least initially, the multiple molecular components of the system. Moreover, the effect observed upon mutating a gene, i.e., on the disruption of the general physiology of a neuron, is not always obviously deducible from the gene product itself. Also, gene function may not always be cell autonomous, therefore making a conclusion regarding gene function and locus of action difficult to draw. Additionally, given the complexity of the spatiotemporal regulation of expression of a majority of genes, with the passage of time, some overview studies have led to the suggestion that behavioral phenotypic specificity is caused by the specificity of the mutation rather than being a real "behavioral gene" (Pflugfelder, 1998). Indeed, of several previously identified "behavioral gene" mutants, mutations in their regulatory sequence and concomitantly on their spatiotemporal patterns of expression rather than a specific mutation in their encoding part itself have proven to be the underlying cause of the phenotype (Pflugfelder, 1998). Thus, the behavioral interpretation of the concepts "clone and sequence" and "behavioral gene" has been revealed to be generally reliant on the limit of pleiotropy of the expressed gene (Hall, 1994b; Pflugfelder, 1998) (in other words, where and when the gene is expressed). In this manner, the toxigenetic method, taking advantage of the enhancer-trap P[GAL4] system, which is based on the trapping of enhancers and silencers for the spatiotemporal modulation of expression of the effector gene, appears to resemble this concept of the pleiotropy of "behavioral genes." In this way, the toxigenetics system and the use of tetanus toxin seem to be judicious and complementary tools for the genetic and molecular characterization of genes to unravel the neural bases of behavior.

Acknowledgments

We are grateful to Reinhart Stocker, François Rouyer, Jean-Marc Devaud, Henrike Scholz, Ulrike Heberlein, and Alberto Ferrus for kindly sharing their unpublished results. We also are grateful to Martin Heisenberg, Jean-François Ferveur, Henrike Scholz, and Doug Guarnieri for the critical reading of the manuscript, and Jeffrey C. Hall for soliciting this review. S. T. Sweeney is currently supported by a Wellcome Trust (U.K.) Prize Travelling Fellowship, A. Keller by the DFG Graduiertenkolleg "Arthropodenverhalten" in Würzburg, Germany, and J. R. Martin by Pasteur Institute, Paris, and by the CNRS, France.

References

Accardo, A., Affinito, M., Carrozzin, M., and Bouquet, F. (1997). Use of the fractal dimension for the analysis of electroencephalographic time series. *Biol. Cybern.* **77,** 339–350.

Ahnert-Hilger, G., Bader, M. F., Bhakdi, S., and Gratzl, M. (1989). Introduction of macromolecules into bovine adrenal medullary chromaffin cells and rat pheochromocytoma cells (PC12) by permeabilization with streptolysin O: Inhibitory effect of tetanus toxin on catecholamine secretion. *J. Neurochem.* **52,** 1751–1758.

Alcorta, E. (1991). Characterization of the electroantennogram in *Drosophila melanogaster* and its use for identifying olfactory capture and transduction mutants. *J. Neurophysiol.* **65,** 702–714.

Allen, M. J., Shan, X., Caruccio, P., Froggett, S. J., Moffat, K. G., and Murphey, R. K. (1999). Targeted expression of truncated *glued* disrupts giant fiber synapse formation in *Drosophila. J. Neurosci.* **19,** 9374–9384.

Archer, B. T., Özçelik, T., Jahn, R., Francke, U., and Südhof, T. C. (1990). Structures and chromosomal localizations of two human genes encoding synaptobrevins 1 and 2. *J. Biol. Chem.* **265,** 17267–17273.

Armstrong, J. D., de Belle, S., Wang, Z., and Kaiser, K. (1998). Metamorphosis of the mushroom bodies; large-scale rearrangements of the neural substrates for associative learning and memory in *Drosophila. Learning Memory* **5,** 102–114.

Baines, R. A., and Bate, M. (1998). Electrophysiological development of central neurons in the *Drosophila* embryo. *J. Neurosci.* **18,** 4673–4683.

Baines, R. A., Robinson, S. G., Fujioka, M., Jaynes, J. B., and Bate, M. (1999). Postsynaptic expression of tetanus toxin light chain blocks synaptogenesis in *Drosophila. Curr. Biol.* **9,** 1267–1270.

Bainton, R. J., Tsai, L. T.-Y., Singh, C. M., Moore, M. S., Neckameyer, W. S., and Heberlein, U. (2000). Dopamine modulates acute responses to cocaine, nicotine and ethanol in *Drosophila. Curr. Biol.* **10,** 187–194.

Bassler, U. (1993). The femur-tibia control system of stick insects—A model system for the study of the neural basis of joint control. *Brain Res. Rev.* **18,** 207–226.

Baumert, M., Maycox, P. R., Navone, F., DeCamilli, P., and Jahn, R. (1989). Synaptobrevin: An integral membrane protein of 18,000 daltons present in small synaptic vesicles of rat brain. *EMBO J.* **8,** 379–384.

Bausenwein, B., Müller, N. R., and Heisenberg, M. (1994). Behavior-dependent activity labeling in the central complex of *Drosophila* during controlled visual stimulation. *J. Comp. Neurol.* **340,** 255–268.

Bausenwein, B., Wolf, R., and Heisenberg, M. (1986). Genetic dissection of optomotor behavior in *Drosophila melanogaster* studies on wild-type and the mutant optomotor-blind/H31. *J. Neurogenet.* **3,** 87–109.

Bellen, H. J., O'Kane, C. J., Wilson, C., Grossniklaus, U., Pearson, R. K., and Gehring, W. J. (1989). P-element mediated enhancer detection: A versatile method to study development in *Drosophila. Genes Dev.* **3,** 1288–1300.

Bello, B., Resendez-Perez, D., and Gehring, W. J. (1998). Spatial and temporal targeting of gene expression in *Drosophila* by means of a tetracycline-dependent transactivator system. *Development* **125,** 2193–2202.

Bier, E., Vaessin, H., Shepherd, S., Lee, K., McCall, K., Barbel, S., Ackerman, L., Caretto, R., Uemura, T., and Grell, E. (1989). Searching for pattern and mutation in the *Drosophila* genome with a P-lacZ vector. *Genes Dev.* **3,** 1273–1287.

Bieschke, E. T., Wheeler, J. C., and Tower, J. (1998). Doxycycline-induced transgene expression during *Drosophila* development and aging. *Mol. Gen. Genet.* **258,** 571–579.

Blagburn, J. M., Alexopoulos, H., Davies, J. A., and Bacon, J. P. (1999). Null mutation in *shaking-B* eliminates electrical, but not chemical, synapses in the *Drosophila* giant fibre system: a structural study. *J. Comp. Neurol.* **404,** 449–458.

Blanchardon, E., Grima, B., Klarsfeld, A., Chélot, A., Hardin, P. E., Préat, T., and Rouyer, F. (2001). Defining the role of *Drosophila* lateral neurons in the control of circadian activity and eclosion rhythms by targeted genetic ablation and PERIOD protein overexpression. *Eur. J. Neurosci.* **13,** 871–888.

Boquet, I., Hitier, R., Dumas, M., and Préat, T. (2000). Central brain postembryonic development in *Drosophila*: Implication of genes expressed at the interhemispheric junction. *J. Neurobiol.* **42,** 33–48.

Brand, A. H., and Dormund, E. L. (1995). The GAL4 system as a tool for unravelling the mysteries of the nervous system. *Curr. Opin. Neurobiol.* **5,** 572–578.

Brand, A. H., and Perrimon, N. (1993). Targeted gene expression as a means of altering cell fates and generating dominant phenotypes. *Development* **118,** 401–415.

Breton, S., Nsumu, N. N., Galli, T., Sabolic, I., Smith, P. J., and Brown, D. (2000). Tetanus toxin-mediated cleavage of cellubrevin inhibits proton secretion in the male reproductive tract. *Am. J. Physiol. Renal Physiol.* **278,** F717–F725.

Broadie, K. S., and Bate, M. (1993). Development of the embryonic neuromuscular synapse of *Drosophila melanogaster. J. Neurosci.* **13,** 144–166.

Broadie, K., Prokop, A., Bellen, H. J., O'Kane, C. J., Schulze, K. L., and Sweeney, S. T. (1995). Syntaxin and synaptobrevin function downstream of vesicle docking in *Drosophila. Neuron* **15,** 663–673.

Brunner, A., Wolf, R., Pflugfelder, G. O., Poeck, B., and Heisenberg, M. (1992). Mutations in the proximal region of the optomotor-blind locus of *Drosophila melanogaster* reveal a gradient of neuroanatomical and behavioral phenotypes. *J. Neurogenet.* **8,** 43–55.

Bryant, P. J. (1978). Pattern formation in imaginal discs. *In* "The Genetics and Biology of Drosophila 2c" (M. Ashburner and T. R. F. Wright, eds.). Academic Press, London.

Budnik, V., Zhong, Y., and Wu, C.-F. (1990). Morphological plasticity of motor axons in *Drosophila* mutants with altered excitability. *J. Neurosci.* **10,** 3754–3768.

Burrows, M. (1987). Parallel processing of proprioceptive signals by spiking local interneurons and motoneurons in the locust. *J. Neurosci.* **7,** 1064–1080.

Cajal, S. R., and Sanchez, D. (1915). Contribucion al conocimiento de los centros nerviosos de los insectos. *Trab. Lab. Invest. Biol. Univ. Madrid* **13**(1).

Carlson, J. R. (1996). Olfaction in *Drosophila:* From odor to behavior. *Trends Genet.* **12,** 175–180.

Chin, A. C., Burgess, R. W., Wong, B. R., Schwarz, T. L., and Scheller, R. H. (1993). Differential expression of transcripts from syb, a *Drosophila melanogaster* gene encoding VAMP (synaptobrevin) that is abundant in non-neuronal cells. *Gene* **131,** 175–181.

Christensen, T. A., Waldrop, B. R., Harrow, I. D., and Hildebrand, J. G. (1993). Local interneurons and information processing in the olfactory glomeruli of the moth *Manduca sexta. J. Comp. Physiol.* **173,** 385–399.

Cole, B. J. (1995). Fractal time in animal behavior: The movement activity of *Drosophila. Anim. Behav.* **50,** 1317–1324.

Connolly, J. B., Roberts, I. J. H., Armstrong, J. D., Kaiser, K., Forte, M., Tully, T., and O'Kane, C. J. (1996). Associative learning disrupted by impaired Gs signaling in *Drosophila* mushroom bodies. *Science* **274,** 2104–2107.

Cooley, L., Kelley, R., and Spradling, A. (1988). Insertional mutagenesis of the *Drosophila* genome with single P elements. *Science* **239,** 1121–1128.

Coombe, P. E. (1986). The large monopolar cells L1 and L2 are responsible for ERG transients in *Drosophila. J. Comp. Physiol.* **159,** 655–665.

Coombe, P. E., and Heisenberg, M. (1986). The structural brain mutant Vacuolar medulla of *Drosophila melanogaster* with specific behavioral defects and cell degeneration in the adult. *J. Neurogenet.* **3,** 135–158.

Crittenden, J. R., Skoulakis, E. M. C., Han, K. A., Kalderon, D., and Davis, R. (1998). Tripartite mushroom body architecture revealed by antigenic markers. *Learning Memory* **5,** 38–51.

Davis, G. W., DiAntonio, A., Petersen, S. A., and Goodman, C. S. (1998). Postsynaptic PKA controls quantal size and reveals a retrograde signal that regulates presynaptic transmitter release in *Drosophila. Neuron* **20,** 305–315.

Davis, R. L. (1993). Mushroom bodies and *Drosophila* learning. *Neuron* **11,** 1–14.

Davis, R. L. (1996). Physiology and biochemistry of *Drosophila* learning mutants. *Physiol. Rev.* **76,** 299–317.

De Belle, J. S., and Heisenberg, M. (1994). Associative odor learning in *Drosophila* abolished by chemical ablation of Mushroom bodies. *Science* **263,** 692–695.

De Belle, J. S., Wulf, J., and Helfrich-Förster, C. (1996). Mushroom bodies do not mediate circadian locomotor activities in Drosophila. In "VIth European Symposium on Drosophila Neurobiology," p. 48.

Deitcher, D. L., Ueda, A., Stewart, B. A., Burgess, R. W., Kidokoro, Y., and Schwarz, T. L. (1998). Distinct requirements for evoked and spontaneous release of neurotransmitter are revealed by mutations in the Drosophila gene neuronal-synaptobrevin. J. Neurosci. **18**, 2028–2039.

Dethier, V. G. (1976). "The Hungry Fly." Harvard Univ. Press, Cambridge, MA.

deZazzo, J., and Tully, T. (1995). Dissection of memory formation: From behavioral pharmacology to molecular genetics. Trends Neurosci **18**, 212–218.

DiAntonio, A., Burgess, R. W., Chin, A. C., Deitcher, D. L., Scheller, R. H., and Schwarz, T. L. (1993). Identification and characterization of Drosophila genes for synaptic vesicle proteins. J. Neurosci. **13**, 4924–4935.

Dickinson, M. H. (1990). Comparison of encoding properties of campaniform sensilla on the fly wing. J. Exp. Biol. **151**, 245–262.

Douglass, J. K., and Strausfeld, N. J. (1996). Visual motion-detection circuits in flies: Parallel direction- and non-direction-sensitive pathways between the medulla and lobula plate. J. Neurosci. **16**, 4551–4562.

Drain, P., Folkers, E., and Quinn, W. G. (1991). cAMP-dependent protein kinase and the disruption of learning in transgenic flies. Neuron **6**, 71–82.

Drescher, W. (1960). Regenerationsversuche am Gehirn von Periplaneta americana unter Berücksichtigung von Verhaltensänderung und Neurosekretion. Z. Morphol. Okol. Tiere **48**, 576–649.

Dunlap, J. C. (1999). Molecular bases for circadian clocks. Cell **96**, 271–290.

Eisel, U., Reynolds, K., Riddick, M., Zimmer, A., Niemann, H., and Zimmer, A. (1993). Tetanus toxin light chain expression in Sertoli cells of transgenic mice causes alterations of the actin cytoskeleton and disrupts spermatogenesis. EMBO J. **12**, 3365–3372.

Elbert, T., Ray, W. J., Kowalik, Z. J., Skinner, J. E., Graf, K. E., and Birbaumer, N. (1994). Chaos and physiology: Deterministic chaos in excitable cell assemblies. Physiol. Rev. **74**, 1–47.

Elferink, L. A., Trimble, W. S., and Scheller, R. H. (1989). Two vesicle-associated membrane protein genes are differentially expressed in the rat central nervous system. J. Biol. Chem **264**, 11061–11064.

Emery, P., Stanewsky, R., Helfrich-Foerster, C., Emery-Le, M., Hall, J. C., and Rosbash, M. (2000). Drosophila CRY is a deep brain circadian photoreceptor. Neuron **26**, 493–504.

Erber, J., Homberg, U., and Gronenberg, W. (1987). Functional roles of the mushroom bodies in insects. In "Arthropod Brain, Its Evolution, Development, Structure and Functions" (A. P. Gupta, ed.), pp. 485–511. Wiley, New York.

Erber, J., Masuhr, T., and Menzel, R. (1980). Localization of short-term memory in the brain of the bee Apis mellifera. Physiol. Entomol. **5**, 343–358.

Ewer, J., Frisch, B., Hamblen-Coyle, M. J., Rosbash, M., and Hall, J. C. (1992). Expression of the period clock gene within different cell types in the brain of Drosophila adults and mosaic analysis of these cells' influence on circadian behavioral rhythms. J. Neurosci. **12**, 3321–3349.

Ferveur, J. F., and Greenspan, R. J. (1998). Courtship behavior of brain mosaics in Drosophila. J. Neurogenet. **12**, 205–226.

Ferveur, J. F., Savarit, F., O'Kane, C. J., Sureau, G., Greenspan, R. J., and Jallon, J. M. (1997). Genetic feminization of pheromones and its behavioral consequences in Drosophila males. Science **276**, 1555–1558.

Ferveur, J. F., Störtkuhl, K. F., Stocker, R. F., and Greenspan, R. J. (1995). Genetic feminization of brain structures and changed sexual orientation in male Drosophila. Science **267**, 902–905.

Finnerty, G. T., and Jefferys, J. G. (2000). 9–16 Hz oscillation precedes secondary generalization of seizure in the rat tetanus toxin model of epilepsy. J. Neurophysiol. **83**, 2217–2226.

Fischbach, K. F., and Dittrich, A. P. M. (1989). The optic lobe of *Drosophila melanogaster*. *Cell Tissue Res.* **258**, 441–476.

Freeman, M. (1996). Reiterative use of the EGF receptor triggers differentiation of all cell types in the *Drosophila* eye. *Cell* **87**, 651–660.

Frisch, B., Hardin, P. E., Hamblen-Coyle, M. J., Rosbash, M., and Hall, J. C. (1994). A promoterless *period* gene mediates behavioral rhythmicity and cyclical *per* expression in a restricted subset of the *Drosophila* nervous system. *Neuron* **12**, 555–570.

Fujii, N., Kimura, K., Yokosawa, N., Tsuzuki, K., and Oguma, K. (1992). A zinc-protease specific domain in botulinum and tetanus neurotoxins. *Toxicon* **30**, 1486–1488.

Gailey, D., Jackson, F., and Siegel, R. (1982). Male courtship in *Drosophila:* The conditioned response to immature males and its genetics control. *Genetics* **102**, 771–782.

Galli, T., Chilcote, T., Mundigi, O., Binz, T., Niemann, H., and DeCamilli, P. (1994). Tetanus-toxin-mediated cleavage of cellubrevin impairs exocytosis of transferrin receptor-containing vesicles in CHO cells. *J. Cell Biol.* **125**, 1015–1024.

Ganetzky, B., and Wu, C.-F. (1993). Neurogenetic analysis of potassium currents in *Drosophila:* Synergistic effects on neuromuscular transmission in double mutants. *J. Neurogenet.* **1**, 17–28.

Gatti, S., Ferveur, J. F., and Martin, J. R. (2000). Genetic identification of neurones controlling a sexually dimorphic behavior. *Curr. Biol.* **10**, 667–670.

Gaumer, S., Guénal, S., Brun, S., Théodore, L., and Mignotte, B. (2000). Bcl-2 and Bax mammalian regulators of apoptosis are functional in *Drosophila*. *Cell Death Differ.* **9**, 804–814.

Gerst, J. E., Rodgers, L., Riggs, M., and Wigler, M. (1992). SNC1, a yeast homolog of the synaptic vesicle associated membrane protein/synaptobrevin gene family: Genetic interactions with the RAS and CAP genes. *Proc. Natl. Acad. Sci. USA* **89**, 4338–4342.

Goldberger, A., and West, B. (1987). Chaos in physiology. *In* "Chaos in Biological Systems" (A. V. Holden, H. Degn, and L. F. Olsen, eds.), pp. 1–5. Plenum Press, New York.

Greenspan, R. J. (1995). Understanding the genetic construction of behavior. *Sci. Am.*, April.

Hall, J. C. (1977). Portions of the central nervous system controlling reproductive behavior in *Drosophila melanogaster*. *Behav. Genet.* **7**, 291–312.

Hall, J. C. (1979). Control of male reproductive behavior by the central nervous system of *Drosophila:* Dissection of a courtship pathway by genetic mosaics. *Genetics* **92**, 437–457.

Hall, J. C. (1982). Genetics of the nervous system in *Drosophila*. *Q. Rev. Biophys.* **15**, 223–479.

Hall, J. C. (1990). Genetics of circadians rhythms. *Annu. Rev. Genet.* **24**, 659–697.

Hall, J. C. (1994a). The mating of a fly. *Science* **264**, 1702–1714.

Hall, J. C. (1994b). Pleiotropy of behavioral genes. *In* "Flexibility and Constraint in behavioral Systems" (R. L. Greenspan, and C. P. Kyriacou, eds.), pp. 15–27. Wiley, New York.

Hall, J. C. (1995). Tripping along the trail to the molecular mechanisms of biological clocks. *Trends Neurosci.* **18**, 230–240.

Hall, J. C., and Greenspan, R. J. (1979). Genetic analysis of *Drosophila* neurobiology. *Annu. Rev. Genet.* **13**, 127–195.

Hall, Z. W., and Sanes, J. R. (1993). Synaptic structure and development: the neuromuscular junction. *Cell* **72**/*Neuron* **10**(suppl.), 99–121.

Han, P. L., Levin, L. R., Reed, R. R., and Davis, R. L. (1992). Preferential expression of the *Drosophila rutabaga* gene in mushroom bodies, neural centers for learning in insects. *Neuron* **9**, 619–627.

Hanesch, U., Fischbach, K. F., and Heisenberg, M. (1989). Neuronal architecture of the central complex in *Drosophila melanogaster*. *Cell Tissue Res.* **257**, 343–366.

Hausen, K. (1981). Monocular and binocular computation of motion in the lobula plate of the fly. *Verh. Dtsch. Zool. Ges.*, 49–70.

Hausen, K., and Wehrhahn, C. (1990). Neural circuits mediating visual flight control in flies. II. Separation of two control systems by microsurgical brain lesions. *J. Neurosci.* **10**, 351–360.

Heimbeck, G., Bugnon, V., Gendre, N., Haberlin, C., and Stocker, R. F. (1999). Smell and taste perception in *Drosophila melanogaster* larva: Toxin expression studies in chemosensory neurons. *J. Neurosci.* **19,** 6599–6609.

Heimbeck, G., Bugnon, V., Gendre, N., and Stocker, R. F. (2000). Brain structures important for chemosensory and sexual behavior of flies. *In* "8th European Symposium on *Drosophila* Neurobiology, Alicante, Spain," p. 53.

Heisenberg, M. (1989). Genetic approach to learning and memory (mnemogenetics) in *Drosophila melanogaster*. *In* "Fundamentals of Memory Formation: Neuronal Plasticity and Brain Function" (H. Rahmann, ed.), pp. 3–45. Fisher Verlag, Stuttgart.

Heisenberg, M. (1994). Central brain function in insects: genetics studies on the mushroom bodies and central complex in *Drosophila*. *In* "Fortschritte der Zoologie: Neural Basis of Behavioral Adaptations" (W. Rathmayer, ed.), pp. 61–79. Fisher, Stuttgart.

Heisenberg, M. (1998). What do the mushroom bodies do for the insect brain? An introduction. *Learning Memory* **5,** 1–10.

Heisenberg, M., Borst, A., Wagner, S., and Byers, D. (1985). *Drosophila* mushroom body mutants are deficient in olfactory learning. *J. Neurogenet.* **2,** 1–30.

Heisenberg, M., and Wolf, R. (1984). "Vision in *Drosophila*: Genetics of Microbehavior." Springer Verlag, Berlin.

Heisenberg, M., Wonneberger, R., and Wolf, R. (1978). Optomotor-blind H31/a *Drosophila* mutant of the lobula plate giant neurons. *J. Comp. Physiol.* **124,** 287–296.

Helfrich-Förster, C. (1995). The period clock gene is expressed in central nervous system neurons which also produce a neuropeptide that reveals the projections of circadians pacemaker cells within the brain of *Drosophila melanogaster*. *Proc. Natl. Acad. Sci. USA* **92,** 612–616.

Helfrich-Förster, C. (1997). Development of pigment-dispersing hormone immunoreactive neurons in the nervous system of *Drosophila melanogaster*. *J. Comp. Neurol.* **380,** 335–354.

Helfrich-Förster, C. (1998). Robust circadian rhythmicity of *Drosophila melanogaster* requires the presence of lateral neurons: A brain-behavioral study of disconnected mutants. *J. Comp. Physiol. A* **182,** 435–453.

Helfrich-Förster, C., and Homberg, U. (1993). Pigment-dispersing hormone immunoreactive neurons in the nervous system of wild-type *Drosophila melanogaster* and several mutants with altered circadian rhythmicity. *J. Comp. Neurol.* **337,** 177–190.

Helfrich-Förster, C., Stengl, M., and Homberg, U. (1998). Organization of the circadian system in insects. *Chronobiol. Int.* **15,** 567–594.

Hiesinger, P. R., Reiter, C., Schau, H., and Fischbach, K.-F. (1999). Neuropil pattern formation and regulation of cell adhesion molecules in *Drosophila* optic lobe development depend on synaptobrevin. *J. Neurosci.* **19,** 7548–7556.

Homberg, U. (1987). Structure and functions of the Central Complex in insects. *In* "Arthropod Brain, Its Evolution, Development, Structure and Functions" (A. P. Gupta, ed.), pp. 347–367. Wiley, New York.

Hotta, Y., and Benzer, S. (1972). Mapping of behavior in *Drosophila* mosaics. *Nature* **240,** 527–535.

Hotta, Y., and Benzer, S. (1976). Courtship in *Drosophila* mosaics: Sex-specific foci for sequential action patterns. *Proc. Natl. Acad. Sci. USA* **73,** 4154–4158.

Huber, F. (1955). Sitz und Bedeutung nervöser Zentren für Instinkthandlungen beim Männchen von *Gryllus campestris* L. *Z. Tierpsychol.* **12,** 12–48.

Huber, F. (1960). Untersuchungen über die Funktion des Zentralnervensystems und insbesondere des Gehirnes bei der Fortbewegung und der Lauterzeugung der Grillen. *Z. vergl. Physio.* **44,** 60–132.

Huber, F. (1963). The role of the central nervous system in Orthoptera during the co-ordination and control of stridulation. *In* "Acoustic Behavior of Animals" (R. G. Busnel, ed.), pp. 440–488. Elsevier, Amsterdam.

Huber, F. (1965). Brain controlled behavior in Orthopterans. In "The Physiology of the Insect Central Nervous System" (J. E. Treherne, and J. W. L. Beament, eds.), pp. 233–246. Academic Press, London/New York.

Hunt, J. M., Bommert, K., Charlton, M. P., Kistner, A., Habermann, E., Augustine, G. J., and Betz, H. (1994). A postdocking role for synaptobrevin in synaptic vesicle fusion. Neuron 12, 1269–1279.

Ilius, M., Wolf, R., and Heisenberg, M. (1994). The central complex of Drosophila melanogaster is involved in flight control: Studies on mutants and mosaics of the gene ellipsoid-body-open. J. Neurogenet. 9, 189–206.

Ito, K., Awano, W., Suzuki, K., Hiromi, Y., and Yamamoto, D. (1997). The Drosophila mushroom body is a quadruple structure of clonal units each of which contains a virtually identical set of neurones and glial cells. Development 124, 761–771.

Ito, K., Suzuki, K., Estes, P., Ramaswami, M., Yamamoto, D., and Strausfeld, N. J. (1998). The organisation of extrinsic neurons and their implications in the functional roles of the mushroom bodies in Drosophila melanogaster Meigen. Learning Memory 5, 52–77.

Jan, Y. N., and Jan, L. Y. (1993). The peripheral nervous system. In "The Development of Drosophila melanogaster" (M. Bate and A. Martinez Arias, eds.), Cold Spring Harbor Laboratory Press, Cold Spring Harbor, NY.

Jarecki, J., Johnson, E., and Krasnow, M. A. (1999). Oxygen regulation of airway branching in Drosophila is mediated by branchless FGF. Cell 99, 211–220.

Jongeneel, C. V., Bouvier, J., and Bairoch, A. (1989). A unique signature identifies a family of zinc-dependent metallopeptidases. FEBS Lett. 312, 110–114.

Kalivas, P. W., Duffy, P., DuMars, L. A., and Skinner, C. (1988). Behavioral and neurochemical effects of acute and daily cocaine administration in rats. J. Pharmacol. Exp. Ther. 245, 485–492.

Kalivas, P. W., Pierce, R. C., Cornish, J., and Sorg, B. A. (1998). A role for sensitization in craving and relapse in cocaine addiction. J. Psychopharmacol. 12, 49–53.

Kaneko, M. (1998). Neural substrates of Drosophila rhythms revealed by mutants and molecular manipulations. Curr. Opin. Neurobiol. 8, 652–658.

Kaneko, M., and Hall, J. C. (2000). Neuroanatomy of cells expressing clock genes in Drosophila: transgenic manipulation of the period and timeless genes to mark the perikarya of circadian pacemaker neurons and their projections. J. Comp. Neurol. 422, 66–94.

Kaneko, M., Helfrich-Förster, C., and Hall, J. C. (1997). Spatial and temporal expression of the period and timeless genes in the developing nervous system of Drosophila: Newly identified pacemaker candidates and novel features of clock gene product cycling. J. Neurosci. 17, 6745–6760.

Kaneko, M., Park, J. H., Cheng, Y., Hardin, P., and Hall, J. C. (2000). Disruption of synaptic transmission or clock gene product oscillations in circadian pacemaker cells of Drosophila cause abnormal behavioral rhythms. J. Neurobiol. 43, 207–203.

Kankel, D. R., and Hall, J. C. (1976). Fate mapping of nervous system and other internal tissues in genetic mosaics of Drosophila melanogaster. Dev. Biol. 48, 1–24.

Kankel, D. R., Ferrus, A., Garen, S. H., Harte, P. J., and Lewis, P. E. (1980). The structure and development of the nervous system. In "The Genetics and Biology of Drosophila 2d" (M. Ashburner, and T. R. F. Wright, eds.). Academic Press, London.

Keller, A., and Heisenberg, M. (1998). Behavioral consequences of tetanus toxin expression in retinal cells. In "VIIth European Symposium on Drosophila Neurobiology," p. 11.

Keller, A., Sweeney, S. T., O'Kane, C. J., and Heisenberg, M. (1999). Tetanus toxin as a tool to study visual processing. Proc. 27th Göttingen Neurobiol. Conf., 446.

Keller, A., Sweeney, S. T., Zars, T., O'Kane, C. J., and Heisenberg, M. Targeted expression of tetanus neurotoxin interferes with behavioral responses in Drosophila. J. Neurobiol. (in press).

Kernan, M. J., Kuroda, M. I., Kreber, R., Baker, B. S., and Ganetzky, B. (1991). nap^ts, a mutation affecting sodium channel activity in Drosophila, is an allele of mele, a regulator of X chromosome transcription. Cell 66, 949–959.

Kerscher, S., Albert, S., Wucherpfennig, D., Heisenberg, M., and Schneuwly, S. (1995). Molecular and genetic analysis of the *Drosophila* mas-1 (mannosidase-1) gene which encodes a glycoprotein processing α1,2-mannosidase. *Dev. Biol.* **168**, 613–626.

King, D. G., and Wyman, R. J. (1980). Anatomy of the giant fibre pathway in *Drosophila*. I. Three thoracic components of the pathway. *J. Neurocytol.* **9**, 753–770.

Konopka, R. J., and Benzer, S. (1971). Clock mutants of *Drosophila melanogaster. Proc. Natl. Acad. Sci. USA* **68**, 2112–2116.

Kyriacou, C. P., and Hall, J. C. (1984). Learning and memory mutations impair acoustic priming of mating behavior in *Drosophila. Nature* **308**, 62–65.

Kyriacou, C. P., and Hall, J. C. (1994). Genetic and molecular analysis of *Drosophila* behavior. *Adv. Genet.* **31**, 139–186.

Laissue, P. P., Reiter, C., Hiesinger, P. R., Halter, S., Fischbach, K. F., and Stocker, R. F. (1999). Three-dimensional reconstruction of the antennal lobe in *Drosophila melanogaster. J. Comp. Neurol.* **405**, 543–552.

Le, A. D., Khanna, J. M.., Kalant, H., and LeBlanc, A. E. (1981). The effect of lesions in the doral, median, and magnus raphe nuclei on the development of tolerance to ethanol. *J. Pharmacol. Exp. Ther.* **218**, 525–9.

Lee, T., Lee, A., and Luo, L. (1999). Development of the *Drosophila* mushroom bodies: sequential generation of three distinct types of neurons from a neuroblast. *Development* **126**, 4065–4076.

Leng, S., and Strauss, R. (1996). A new walking impaired *Drosophila* mutant has a structural defect in the protocerebral bridge of the central complex. *Proc. 24th Göttingen Neurobiol. Conf.*, 134.

Leng, S., and Strauss, R. (1997). Impaired step lengths common to three unrelated *Drosophila* mutant lines with common brain defects confirm the involvement of the protocerebral bridge in optimizing walking speed. *Proc. 25th Göttingen Neurobiol. Conf.*, 294.

Le Van Quyen, M., Martinerie, J., Baulac, M., and Varela, F. (1999). Anticipating epileptic seizures in real time by a non-linear analysis of similarity between EEG recording. *NeuroReport* **10**, 2149–2155.

Li, H, Chaney, S., Forte, M., and Hirsh, J. (2000). Ectopic G-protein expression in dopamine and serotonin neurons blocks cocaine sensitization in *Drosophila melanogaster. Curr. Biol.* **10**, 211–214.

Li, L., Wolf, R., Ernst, R., and Heisenberg, M. (1999). Context generalization in *Drosophila* visual learning requires the mushroom bodies. *Nature* **400**, 753–756.

Li, Y., Foran, P., Fairweather, N. F., dePaiva, A., Weller, U., Dougan, G., and Dolly, J. O. (1994). A single mutation in the recombinant light chain of tetanus toxin abolishes its proteolytic activity and removes the toxicity seen after reconstitution with native heavy chain. *Biochemistry* **33**, 7014–7020.

Lin, D. M., Auld, V. J., and Goodman, C. S. (1995). Targeted neuronal cell ablation in the *Drosophila* embryo: Pathfinding by follower growth cones in the absence of pioneers. *Neuron* **14**, 707–715.

Link, E., Edelman, L., Chou, J. H., Binz, T., Yamasaki, S., Eisel, U., Baumert, M., Südhof, T. C., Niemann, H., and Jahn, R. (1992). Tetanus toxin action: Inhibition of neurotransmitter release linked to synaptobrevin proteolysis. *Biochem. Biophys. Res. Commun.* **189**, 1017–1023.

Lloyd, T. E., Verstreken, P., Ostrin, E. J., Phillippi, A., Lichtarge, O., and Bellen, H. J. (2000). A genome-wide search for synaptic vesicle cycle proteins in *Drosophila. Neuron* **26**, 45–50.

Lutzenberger, W., Preissl, H., and Pulvermüller, F. (1995). Fractal dimension of electroencephalographic time series and underlying brain processes. *Biol. Cybern.* **73**, 477–482.

MacLeod, K., and Laurent, G. (1996). Distinct mechanisms for synchronization and temporal patterning of odor-encoding neural assemblies. *Science* **274**, 976–979.

Manning, A. (1967). Antennae and sexual receptivity in *Drosophila melanogaster* females. *Science* **158**, 136–137.

Martin, J. R., Ernst, R., and Heisenberg, M. (1998). Mushroom bodies suppress locomotor activity in *Drosophila melanogaster. Learing Memory* **5**, 179–191.

Martin, J. R., Ernst, R., and Heisenberg, M. (1999a). Temporal pattern of locomotor activity in *Drosophila melanogaster. J. Comp. Physiol. A* **184**, 73–84.

Martin, J. R., Raabe, T., and Heisenberg, M. (1999b). Central complex substructures are required for the maintenance of locomotor activity in *Drosophila melanogaster*. *J. Comp. Physiol. A* **185,** 277–288.

Martin, J. R., Faure, P., and Ernst, R. The power law distribution for walking-time intervals correlates with the ellipsoid-body in *Drosophila*. *J. Neurogenet.* (in press).

Martinerie, J., Adam, C., Le Van Quyen, M., Baulac, M., Clemenceau, S., Renault, B., and Varela, F. J. (1998). Epileptic seizures can be anticipated by non-linear analysis. *Nat. Med.* **4,** 1173–1176.

Mayford, M., Bach, M. E., Huang, Y. Y., Wang, L., Hawkins, R. D., and Kandel, E. R. (1996). Control of memory formation through regulated expression of CaMKII transgene. *Science* **274,** 1678–1683.

McBride, S. M. J., Giuliani, G., Choi, C., Krause, P., Correale, D., Watson, K., Baker, G., and Siwicki, K. K. (1999). Mushroom body ablation impairs short-term memory and long-term memory of courtship conditioning in *Drosophila melanogaster*. *Neuron* **24,** 967–977.

McCall, K, and Steller, H. (1997). Facing death in the fly: Genetic analysis of apoptosis in *Drosophila*. *Trends Genet.* **13,** 222–226.

McClung, C., and Hirsh, J. (1998). Stereotypic behavioral responses to free-base cocaine and the development of behavioral sensitization in *Drosophila melanogaster*. *Curr. Biol.* **8,** 109–112.

McClung, C., and Hirsh, J. (1999). The trace amine tyramine is essential for sensitization to cocaine in *Drosophila*. *Curr. Biol.* **9,** 853–860.

McKenna, M., Monte, P., Helfand, S. L., Woodward, C., and Carlson, J. (1989). A simple chemosensory response in *Drosophila* and the isolation of *acj* mutants in which it is affected. *Proc. Natl. Acad. Sci. USA* **86,** 8118–8122.

McMahon, H. T., Ushkaryov, Y. A., Edelmann, L., Link, E., Binz, T., Niemann, H., Jahn, R., and Südhof, T. C. (1993). Cellubrevin is a ubiquitous tetanus-toxin substrate homologous to a putative synaptic vesicle fusion protein. *Nature* **364,** 346–349.

McNabb, S. L., Baker, J. D., Agapite, J., Steller, H., Riddiford, L. M., and Truman, J. W. (1997). Disruption of a behavioral sequence by targeted death of peptidergic neurons in *Drosophila*. *Neuron* **19,** 813–823.

McRobert, S. P., and Tompkins, L. (1983). Courtship of young males is ubiquitous in *Drosophila melanogaster*. *Behav. Genet.* **13,** 517–523.

Meinertzhagen, I. A., and O'Neil, S. D. (1991). Synaptic organization of columnar elements in the lamina of the wild type in *Drosophila melanogaster*. *J. Comp. Neurol.* **305,** 232–263.

Mellanby, J., Johansen-Berg, H., Leyland, R., and Milward, A. J. (1999). The effect of tetanus toxin-induced limbic epilepsy on the exploratory response to novelty in the rat. *Epilepsia* **40,** 1058–1061.

Menzel, R., Erber, J., and Masuhr, T. (1974). Learning and memory in the honeybee. *In* "Experimental Analysis of Insect Behavior" (L. B. Browne, ed.), pp. 195–217. Springer, Berlin.

Miller, A. (1950). The internal anatomy and histology of the imago of *Drosophila melanogaster*. *In* "Biology of *Drosophila*" (M. Demerec, ed.), pp. 420–534. Hafner, New York.

Mochida, S., Poulain, B., Weller, U., Habermann, E., and Tauc, L. (1989). Light chain of tetanus toxin intracellularly inhibits acetylcholine release at neuro-neuronal synapses, and its internalisation is mediated by heavy chain. *FEBS Lett.* **253,** 47–51.

Mochida, S., Poulain, B., Eisel, U., Binz, T., Kurazono, H., Niemann, H., and Tauc, L. (1990). Exogenous mRNA encoding tetanus or botulinum neurotoxins expressed in *Aplysia* neurons. *Proc. Natl. Acad. Sci. USA* **87,** 7844–7848.

Moore, M. S., DeZazzo, J., Luk, A. Y., Tully, T., Singh, S. M., and Heberlein, U. (1998). Ethanol intoxication in *Drosophila*: Genetic and pharmacological evidence for regulation by the cAMP signaling pathway. *Cell* **93,** 997–1007.

Moses, K., and Rubin, G. M. (1991). Glass encodes a site-specific DNA-binding protein that is regulated in response to positional signals in the developing *Drosophila* eye. *Genes Dev.* **5,** 583–593.

Nalbach, G., and Hengstenberg, R. (1994). The halteres of the blowfly Calliphora. II. Three dimensional organization of compensatory reactions to real and stimulated rotation. *J. Comp. Physiol.* A **175**, 695–708.

Neckameyer, W. S. (1998). Dopamine and mushroom bodies in *Drosophila*: Experience-dependent and -independent aspects of sexual behavior. *Learning Memory* **5**, 157–165.

Nighorn, A., Healy, M. J., and Davis, R. L. (1991). The cyclic AMP phosphodiesterase encoded by the *Drosophila* dunce gene is concentrated in the mushroom body neuropil. *Neuron* **6**, 455–467.

Nishikawa, K., and Kidokoro, Y. (1995). Junctional and extrajunctional glutamate receptor channels in *Drosophila* embryos and larvae. *J. Neurosci.* **15**, 7905–7915.

O'Dell, K. M. C., Armstrong, J. D., Yang, M. Y., and Kaiser, K. (1995). Functional dissection of the *Drosophila* mushroom bodies by selective feminization of genetically defined subcompartments. *Neuron* **15**, 55–61.

O'Kane, C. J., and Gehring, W. J. (1987). Detection in situ of genomic regulatory elements in *Drosophila*. *Proc. Natl. Acad. Sci. USA* **84**, 9123–9127.

O'Kane, C. J., Schiavo, G., and Sweeney, S. T. (1999). Toxins that affect neurotransmitter release. In "Neurotransmitter Release" (H. Bellen, ed.), *Frontiers in Molecular Biology* (B. D. Hames, and D. M. Glover, Series, eds.), pp. 208–236. Oxford Univ. Press, Oxford.

Otsuna, H., and Ito, K. (2000). Comprehensive identification of projection neurones connecting optic lobes and the central brain. In "VIIIth European Symposium on *Drosophila* Neurobiology," p. 99.

Otto, D. (1971). Untersuchungen zur zentralnervösen Kontrolle der Lauterzeugung von Grillen. *Z. Vergl. Physiol.* **74**, 227–271.

Park, J. H., and Hall, J. C. (1998). Isolation and chronobiological analysis of a neuropeptide pigment-dispersing factor gene in *Drosophila melanogaster*. *J. Biol. Rhythms* **13**, 219–228.

Park, J. H., Helfrich-Förster, C., Lee, G., Liu, L., Rosbash, M., and Hall, J. C. (2000). Differential regulation of circadian pacemaker output by separate clock genes in *Drosophila*. *Proc. Natl. Acad. Sci. USA* **97**, 3608–3613.

Penner, R., Neher, E., and Dreyer, F. (1986). Intracellularly injected tetanus toxin inhibits exocytosis in bovine adrenal chromaffin cells. *Nature*, **324**, 76–78.

Peterson, B. S., and Leckman, J. F. (1998). The temporal dynamics of tics in Gilles de la Tourette syndrome. *Biol. Psychiatry* **15**, 1337–1348.

Pflugfelder, G. O. (1998). Genetics lesions in *Drosophila* behavioral mutants. *Behav. Brain Res.* **95**, 3–15.

Pflugfelder, G. O. (1999). Structure–function analysis of the *Drosophila* optic lobes. In "Handbook of Molecular-Genetic Techniques for Brain and Behavior Research: Techniques in the Behavioral and Neural Sciences" (W. E. Crusio, and R. T. Gerlai, eds.), Vol. 13, Elsevier, Amsterdam.

Pflugfelder, G. O., and Heisenberg, M. (1995). Optomotor-blind of *Drosophila melanogaster*: A neurogenetic approach to optic lobe development and optomotor behavior. *Comp. Biochem. Physiol.* **110**, 185–202.

Pflugfelder, G. O., Schwarz, H., Roth, H., Poeck, B., Sigl, A., Kerscher, S., Jonschker, B., Pak, W. L., and Heisenberg, M. (1990). Genetic and molecular characterization of the optomotor-blind gene locus in *Drosophila melanogaster*. *Genetics* **126**, 91–104.

Phelan, P., Stebbings, L. A., Baines, R. A., Bacon, J. P., Davies, J. A., and Ford, C. C. (1998). *Drosophila* shaking-B protein forms gap junctions in paired *Xenopus* oocytes. *Nature* **391**, 181–184.

Post, R. M., and Rose, H. (1976). Increasing effects of repetitive cocaine administration in the rat. *Nature* **260**, 731–732.

Pothos, E. N., Przedborski, S., Davila, V., Schmitz, Y., and Sulzer, D. (1998). D2-like dopamine autoreceptor activation reduces quantal size in PC12 cells. *J. Neurosci.* **8**, 5575–5585.

Poulain, B., Mochida, S., Weller, U., Hogy, B., Habermann, E., Wadsworth, J. D. F., Shone, C. C., Dolly, J. O., and Tauc, L. (1991). Heterologous combinations of heavy and light chains from botulinum toxin A and tetanus toxin inhibit neurotransmitter release in *Aplysia*. *J. Biol. Chem.* **266**, 9580–9585.

Protopopov, V., Govindan, B., Novick, P., and Gerst, J. E. (1993). Homologs of the synaptobrevin/VAMP family of synaptic vesicle proteins function on the late secretory pathway in S. cerevisiae. *Cell* **74**, 855–861.

Quan, F., Thomas, L., and Forte, M. (1991). *Drosophila* stimulatory G protein$_\alpha$ subunit activates mammalian adenylyl cyclase but interacts poorly with mammalian receptors: Implications for receptor-G protein interaction. *Proc. Natl. Acad. Sci. USA* **88**, 1898–1902.

Rajashekhar, K. P., and Singh, R. N. (1994). Organization of motor neurons innervating the proboscis musculature in *Drosophila melanogaster* Meigen, Diptera: Drosophilidae. *Int. J. Insect Morph. Embryol.* **23**, 225–242.

Reddy, S., Jin, P., Trimarchi, J., Caruccio, P., Phillis, R., and Murphey, R. K. (1997). Mutant molecular motors disrupt neural circuits in *Drosophila*. *J. Neurobiol.* **33**, 711–723.

Regazzi, R., Sadoul, K., Meda, P., Kelly, R. B., Halban, P. A., and Wollheim, C. B. (1996). Mutational analysis of VAMP domains implicated in Ca^{2+}-induced insulin exocytosis. *EMBO J.* **15**, 6951–6959.

Renn, S. C. P., Park, J. H., Rosbash, M., Hall, J. C., and Taghert, P. H. (1999a). A pdf neuropeptide gene mutation and ablation of PDF neurons each cause severe abnormalities of behavioral circadian rhythms in *Drosophila*. *Cell* **99**, 791–802.

Renn, S. C., Armstrong, J. D., Yang, M., Wang, Z., An, X., Kaiser, K., and Taghert, P. H. (1999b). Genetic analysis of the *Drosophila* ellipsoid-body neuropil: Organization and development of the central complex. *J. Neurobiol.* **41**, 189–207.

Robertson, H. M., Preston, C. R., Phillis, R. W., Johnsonschlitz, D. M., and Benz, W. K. (1988). A stable genomic source of P-element transposase in *Drosophila*. *Genetics* **118**, 461–470.

Rodrigues, V. (1988). Spatial coding of olfactory information in the antennal lobe of *Drosophila melanogaster*. *Brain Res.* **453**, 299–307.

Rosbash, M., and Hall, J. C. (1989). The molecular biology of circadian rhythms. *Neuron* **3**, 387–398.

Saitoe, M., Tanaka, S., Takata, K., and Kidokoro, Y. (1997). Neural activity affects distribution of glutamate receptors during neuromuscular junction formation in *Drosophila* embryos. *Dev. Biol.* **184**, 48–60.

Sayeed, O., and Benzer, S. (1996). Behavioral genetics of thermosensation and hygrosensation in *Drosophila*. *Proc. Natl. Acad. Sci. USA* **93**, 6079–6084.

Schiavo, G., Benfenati, F., Poulain, B., Rossetto, O., DeLaureto, P. P., DasGupta, B. R., and Montecucco, C. (1992b). Tetanus and botulinum-B neurotoxins block neurotransmitter release by proteolytic cleavage of synaptobrevin. *Nature* **359**, 832–835.

Schiavo, G., Matteoli, M., and Montecucco, C. (2000). Neurotoxins affecting neuroexocytosis. *Physiol. Rev.* **80**, 717–766.

Schiavo, G., Poulain, B., Rossetto, O., Benfenati, F., Tauc, L., and Montecucco, C. (1992a). Tetanus toxin is a zinc protein and its inhibition of neurotransmitter release and protease activity depend on zinc. *EMBO J.* **11**, 3577–3583.

Scholz, H., Ramond, J., Singh, C. M., and Heberlein, U. (2000). Functional ethanol tolerance in *Drosophila*. *Neuron* **28**, 261–271.

Shimada, I., Kawazoe, Y., and Hara, H. (1993). A temporal model of animal behavior based on a fractality in the feeding of *Drosophila melanogaster*. *Biol. Cybern.* **68**, 477–481.

Siegel, R. W., and Hall, J. C. (1979). Conditioned responses in courtship behavior of normal and mutant *Drosophila*. *Proc. Natl. Acad. Sci. USA* **76**, 3430–3434.

Siegel, R. W., Hall, J. C., Gailey, D. A., and Kyriacou, C. P. (1984). Genetic elements of courtship in *Drosophila*: Mosaics and learning mutants. *Behav. Genet.* **14**, 383–410.

Skoulakis, E. M., Kalderon, D., and Davis, R. L. (1993). Preferential expression in mushroom bodies of the catalytic subunit of protein kinase A and its role in learning and memory. *Neuron* **11**, 197–208.

Smith, H. K., Roberts, I. J. H., Allen, M. J., Connolly, J. B., Moffatt, K. G., and O'Kane, C. J. (1996). Inducible ternary control of transgene expression and cell ablation in *Drosophila*. *Dev. Genes Evol.* **206**, 14–24.

Stocker, R. F. (1994). The organization of the chemosensory system in *Drosophila melanogaster*: A review. *Cell Tissue Res.* **275**, 3–26.

Stocker, R. F., Heimbeck, G., Gendre, N., and de Belle, J. S. (1997). Neuroblast ablation in *Drosophila* P[GAL4] lines reveals origins of olfactory interneurons. *J. Neurobiol.* **32**, 443–456.

Stocker, R. F., Singh, R. N., Schorderet, M., and Siddiqi, O. (1983). Projection patterns of different types of antennal sensilla in the antennal glomeruli of *Drosophila melanogaster*. *Cell Tissue Res.* **232**, 237–248.

Strausfeld, N. J. (1976). "Atlas of an Insect Brain." Springer, Heidelberg.

Strauss, R., Hanesch, U., Kinkelin, M., Wolf, R., and Heisenberg, M. (1992). *No-bridge* of *Drosophila melanogaster*: Portrait of a structural brain mutant of the central complex. *J. Neurogenet.* **8**, 125–155.

Strauss, R., and Heisenberg, M. (1990). Coordination of legs during straight walking and turning in *Drosophila melanogaster*. *J. Comp. Physiol.* **167**, 403–412.

Strauss, R., and Heisenberg, M. (1993). A higher control center of locomotor behavior in the *Drosophila* brain. *J. Neurosci.* **13**, 1852–1861.

Südhof, T. C., Baumert, M., Perin, M. S., and Jahn, R. (1989). A synaptic vesicle membrane protein is conserved from mammals to *Drosophila*. *Neuron* **2**, 1475–1481.

Sweeney, S. T. (1996). Targeted expression of tetanus toxin light chain in *Drosophila melanogaster*. Ph.D. thesis, University of Cambridge.

Sweency, S. T., Broadie, K., Keane, J., Niemann, H., and O'Kane, C. J. (1995). Targeted expression of tetanus toxin light chain in *Drosophila* specifically eliminates synaptic transmission and causes behavioral defects. *Neuron* **14**, 341–351.

Sweeney, S. T., Hidalgo, A., deBelle, J. S., and Keshishian, H. (2000). Functional cell ablation. In " *Drosophila* Protocols" (W. Sullivan, M. Ashburner, and R. S. Hawley, eds.). CSHL Press, New York.

Taylor, B. J., Villella, A., Ryner, L. C., Baker, B. S., and Hall, J. C. (1994). Behavioral and neurobiological implications of sex-determining factors in *Drosophila*. *DEv. Genet.* **15**, 275–296.

Tettamanti, M., Armstrong, J. D., Endo, K., Yang, M. Y., Furukubo-Tokunaga, K., Kaiser, K., and Reichert, H. (1997). Early development of the *Drosophila* mushroom bodies, brain centres for associative learning and memory. *Dev. Genes. Evol.* **207**, 242–252.

Theodosiou, N. A., and Xu, T. (1998). Use of the FLP/FRT system to study *Drosophila* development. *Methods* **14**, 355–365.

Tissot, M., Gendre, N., and Stocker, R. F. (1998). *Drosophila* P[Gal4] lines reveal that motor neurons involved in feeding persist through metamorphosis. *J. Neurobiol.* **37**, 237–250.

Trimarchi, J. R., and Murphey, R. K. (1997). The shaking-B^2 mutation disrupts electrical synapses in a flight circuit in adult *Drosophila*. *J. Neurosci.* **17**, 4700–4710.

Trimble, W. S., Cowan, D. M., and Scheller, R. H. (1988). VAMP-1: A synaptic vesicle-associated integral membrane protein. *Proc. Natl. Acad. Sci. USA* **85**, 4538–4542.

Tsien, J. Z., Chen, D. F., Gerber, D., Tom, C., Mercer, E. H., Anderson, D. J., Mayford, M., Kandel, E. R., and Tonegawa, S. (1996). Subregion and cell type-restricted gene knockout in mouse brain cell. *Cell* **87**, 1317–1326.

Vaias, L. J., Napolitano, L. M., and Tompkins, L. (1993). Identification of stimuli that mediate experience-dependent modification of homosexual courtship in *Drosophila melanogaster*. *Behav. Genet.* **23**, 91–97.

van der Kloot, W. G., and Williams, C. M. (1953). Cocoon construction by the *Cecropia* silkworm. *Behavior* **5**, 141–174.

van Roessel, P., and Brand, A. H. (2000). GAL4-mediated ectopic gene expression in *Drosophila*. In "Drosophila Protocols" (W. Sullivan, M. Ashburner, and R. S. Hawley, eds.), pp. 439–447. CSHL Press, New York.

Verhage, M., Maia, A. S., Plomp, J. J., Brussard, A. B., Heeroma, J. H., Vermeer, H., Toonen, R. F., Hammer, R. E., van den Berg, T. K., Missler, M., Geuze, H. J., and Südhof, T. C. (2000). Synaptic assembly of the brain in the absence of neurotransmitter secretion. *Science* **287,** 865–869.

Vowles, D. M. (1964). Models in the insect brain. In "Neural Theory and Modeling" (E. Reriss, ed.). Stanford Univ. Press, Stanford, CA.

Wahdepuhl, M. (1983). Control of grasshopper singing behavior by the brain responses to electrical stimulation. *Z. Tierpsychol.* **63,** 173–200.

Wahdepuhl, M., and Huber, F. (1979). Elicitation of singing and courtship movements by electrical stimulation of the brain of the grasshopper. *Naturwissenschaften* **66,** 320–322.

Wannek, U., and Strauss, R. (1997). Turning strategies of the walking fly, *Drosophila melanogaster*, and impairments thereof in the brain defective mutant *no-bridge*. *Proc. 25th Göttingen Neurobiol. Conf.*, 292.

Wilson, C., Pearson, R. K., Bellen, H. J., O'Kane, C. J., Grossniklaus, U., and Gehring, W. J. (1989). P-element mediated enhancer detection: An efficient method for isolating and characterising developmentally regulated genes in *Drosophila*. *Genes Dev.* **3,** 1301–1313.

Wolf, R., and Heisenberg, M. (1991). Basic organisation of operant behavior as revealed in *Drosophila* flight orientation. *J. Comp. Physiol. A* **169,** 699–705.

Wolf, R., Wittig, T., Liu, L., Wustmann, G., Eyding, D., and Heisenberg, M. (1998). *Drosophila* mushroom bodies are dispensable for visual, tactile, and motor learning. *Learning Memory* **5,** 166–178.

Yamamoto, D., Jallon, J.-M., and Komatsu, A. (1997). Genetic dissection of sexual behavior in *Drosophila melanogaster*. *Annu. Rev. Entomol.* **42,** 551–585.

Yamasaki, S., Baumeister, A., Binz, T., Blasi, J., Link, E., Cornille, F., Roques, B., Südhof, T. C., Jahn, R., and Niemann, H. (1994b). Cleavage of members of the synaptobrevin/VAMP family by types D and F botulinum neurotoxins and tetanus toxin. *J. Biol. Chem* **269,** 12764–12772.

Yamasaki, S., Hu, Y., Binz, T., Kalkuhl, A., Kurazono, H., Tamura, T., Jahn, R., Kandel, E., and Niemann, H. (1994a). Synaptobrevin/vesicle-associated membrane protein (VAMP) *of Aplysia californica:* Structure and proteolysis by tetanus toxin and botulinal neurotoxins type D and F. *Proc. Natl. Acad. Sci. USA.* **91,** 4688–4692.

Yang, M. Y., Armstrong, J. D., Vilinsky, I., Strausfeld, N. J., and Kaiser, K. (1995). Subdivision of the *Drosophila* mushroom bodies by enhancer-trap expression patterns. *Neuron* **15,** 45–54.

Yasuyama, K., and Meinertzhagen, I. A. (1999). Extraretinal photoreceptors at the compound eye's posterior margin in *Drosophila melanogaster*. *J. Comp. Neurol.* **412,** 193–202.

Zars, T., Fischer, M., Schulz, R., and Heisenberg, M. (2000a). Localization of a short-term memory in *Drosophila*. *Science* **288,** 672–675.

Zars, T., Wolf, R., Davis, R., and Heisenberg, M. (2000b). Tissue-specific expression of the type I adenylyl cyclase rescues the *rutabaga* mutant memory defect: in search of the engram. *Learning Memory* **7,** 18–31.

2

Germline Transformants Spreading Out to Many Insect Species

Peter W. Atkinson
Department of Entomology
University of California
Riverside, California 92521

Anthony A. James
Department of Molecular Biology and Biochemistry
University of California
Irvine, California 92697

Advances in Genetics, Vol. 47

ABSTRACT

The past 5 years have witnessed significant advances in our ability to introduce genes into the genomes of insects of medical and agricultural importance. A number of transposable elements now exist that are proving to be sufficiently robust to allow genetic transformation of species within three orders of insects. In particular all of these transposable elements can be used genetically to transform mosquitoes. These developments, together with the use of suitable genes as genetic markers, have enabled several genes and promoters to be transferred between insect species and their effects on the phenotype of the transgenic insect determined. Within a very short period of time, insights into the function of insect promoters in homologous and heterologous insect species are being gained. Furthermore, strategies aimed at ameliorating the harmful effects of pest insects, such as their ability to vector human pathogens, are now being tested in the pest insects themselves. We review the progress that has been made in the development of transgenic technology in pest insect species and conclude that the repertoire of transposable element-based genetic tools, long available to *Drosophila* geneticists, can now be applied to other insect species. In addition, it is likely that these developments will lead to the generation of pest insects that display a significantly reduced ability to transmit pathogens in the near future.   © 2002, Elsevier Science (USA).

I. INTRODUCTION: THE CONTINUING RELEVANCE OF GENETIC TRANSFORMATION TECHNOLOGIES IN THE GENOMICS ERA

Genetic transformation is the process of integrating exogenous DNA into the germline of whole organisms so that it is inherited in subsequent generations. The intent of these efforts is to create a stable change in the phenotype of the target organism that can be used to answer basic questions about the physiological, genetic, or other effects of the integrated DNA. As such, transformation technologies have had a profound effect on our understanding of the molecular genetics of those organisms in which it has been developed. They provide a means by which the identification and function of cloned genes can be validated and directly enables the functional consequences of *in vitro* generated mutations in genes to be evaluated in the organism itself. As genomic technologies become applied to an increasing number of organisms, and whole genome sequencing projects become more commonplace, genetic transformation will be increasingly important in unequivocally establishing the genetic function of newly discovered genes and in establishing the interrelationships between genes and the proteins they encode. The inability to clearly define the function of presumptive genes identified from genomic sequencing projects may become a significant bottleneck in those species in which a robust gene transfer technology is absent.

Fortunately, genetic transformation is now also becoming commonplace in many insect species. Genetic transformation of the vinegar fly, *Drosophila melanogaster*, has been a widely applied technique since 1982; the simplicity of the methods, together with the large number of both *Drosophila* mutant stocks and cloned genes, serves to further enhance genetic analyses of this insect. Within several years of the first report of *D. melanogaster* transformation, sophisticated transformation-based genetic techniques such as gene tagging and enhancer trapping were deployed, enabling genes to be identified and cloned on the basis of their tissue-specific or temporal expression properties. With the entire genome of *D. melanogaster* now sequenced and in excess of 13,000 genes being identified on the basis of gene organization and primary sequence (Adams *et al.*, 2000), genetic transformation will play an important role in assigning function to a large number of these genes. For example, reverse-genetics approaches using gene replacement techniques (Rong and Golic, 2000) or RNA inhibition (Kennerdell and Carthew, 2000) will enable the assessment of the impact of null mutations of specific genes on phenotypes.

A robust transformation technology consists of techniques that allow reasonably high efficiencies of integration of exogenous DNA into a target species. These techniques depend on the availability of efficient marker genes for transformation, and gene vectors that integrate the foreign DNA into the chromosomes. The most successful approaches in insects rely on transposable element-mediated integration of DNA. Until recently, this technology had not been developed in insects other than *D. melanogaster* and closely related drosophilids. This situation changed abruptly in late 1995 with the first demonstration of *Minos* transposable element transformation of the Mediterranean fruit fly, *Ceratitis capitata* (Loukeris *et al.*, 1995a). Since then, there has been an impressive increase in both the number of insect species that have been transformed and the number of transposable element vectors that are capable of being used to transform them (Table 2.1). The number of insect species within each common group of transformed insects, combined with the number of transposable elements used to achieve this, is shown in Figure 2.1. If this same graph had been compiled by early 1995, the only group that would be represented would be the vinegar flies—the family *Drosophilidae*. The progress made in the past 5 years has been remarkable, particularly the development of four separate transformation systems for use in mosquitoes. Six insect-derived transposable elements now exist that can act as gene vectors in insects. Two of these, *P* and *hobo*, appear restricted for use in drosophilids only, while the remaining four, *Hermes*, *mariner* (represented by the *Mos*1 element), *Minos*, and *piggyBac*, have far wider host ranges. All are Class II transposable elements, mobilizing through a DNA intermediate (Finnegan, 1985).

Table 2.1. Genetic Transformation of Insects by Transposase-Mediated Recombination of Transposable Elements[a]

Common name	Family or superfamily	Species transformed	Transposable elements used	References
Fruit flies	Tephritidae	*Ceratitus capitata*	*Minos*	Loukeris *et al.*, 1995a.
			piggyBac	Handler *et al.*, 1998.
			Hermes	Michel *et al.*, 2001.
		Bactrocera dorsalis	*piggyBac*	Handler and McCombs, 2000.
		Anastrapha suspensa	*piggyBac*	Handler and Harrell, 2001.
Vinegar flies	Drosophilidae	*Drosophila melanogaster*	*P*	Rubin and Spradling, 1982.
			hobo	Blackman *et al.*, 1989.
			Mos1	Lidholm *et al.*, 1993.
			Minos	Loukeris *et al*, 1995b.
			Hermes	O'Brochta *et al.*, 1996.
			piggyBac	Handler and Harrell, 1999.
		Drosophila virilis	*hobo*	Lozovskaya *et al.*, 1996.
			Mos1	Lohe and Hartl, 1996.
		Drosophila simulans	*P*	Scavarda and Hartl, 1984.
		Drosophila hawaiiensis	*P*	Brennan *et al.*, 1984.
Houseflies	Muscidae	*Musca domestica*	*piggyBac*	Hediger *et al.*, 2001.
			Mos1	Yoshiyama *et al.*, 2000.
		Stomoxys calcitrans	*Hermes*	Lehane *et al.*, 2000
Blowflies	Calliphoridae	*Lucilia cuprina*	*piggyBac*	Heinrich *et al.*, 2002.
Mosquitoes	Culicidae	*Aedes aegypti*	*Hermes*	Jasinskiene *et al.*, 1998.
			Mos1	Coates *et al.*, 1998.
			piggyBac	Kokoza *et al.*, 2001.
		Anopheles gambiae	*piggyBac*	Grossman *et al.*, in press.
		Anopheles stephensi	*Minos*	Catteruccia *et al.*, 2000.
			piggyBac	M. Jacobs-Lorena, personal comm.
		Anopheles albimanus	*piggyBac*	O. P. Perera, R. Harrell, and A. M. Handler, personal comm.
		Culex quinquefasciatus	*Hermes*	M. L. Allen *et al.*, 2001.
Silkworm moths	Bombycidae	*Bombyx mori*	*piggyBac*	Tamura *et al.*, 2000.
Gelechiid moths	Gelechiidae	*Pectinophoa gossypiella*	*piggyBac*	Peloquin *et al.*, 2000.
Darkling beetles	Tenebrionidae	*Tribolium castaneum*	*piggyBac*	Berghammer *et al.*, 1999.
			Hermes	Berghammer *et al.*, 1999.

[a]Only the initial report of transformation of a given species by a given transposable element is cited.

In this review we briefly summarize the progress that has been made and the questions that genetic transformation now enables us to address in nondrosophilid insect species. Significantly, in only a short period of time, transformation of these insect species has enabled the functional analysis of isolated promoters to be examined, sometimes with unexpected results. Furthermore, the ability to transform insect species that have, or will, become the foci of mapping and genomic projects will provide researchers with the means to directly address the function of unknown genes that have been positionally cloned. Finally,

TE transformation of insects

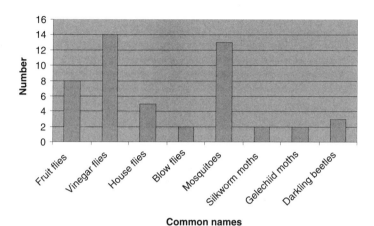

Common names

Figure 2.1. Graphical representation of the numbers of insect species, classified into their common groups, transformed by transposable elements. Each column with each group represents a compilation of the number of species transformed plus the number of transposable elements used. Thus *P* element transformation (1 transposable element) of *D. melanogaster* (1 species) counts as a value of 2 in the vinegar fly group.

while *D. melanogaster* serves as an excellent model in which to perform genetic dissections of complex aspects of development, neurobiology, and other basic science areas, some insect-related biology such as hematophagy, diapause, and host-seeking, are better studied in other species. The application of transformation technology to these questions and to applied goals involving medically and agriculturally important insects widens the research scope, providing novel ways to solve problems arising from the negative impact of insects on human health and society.

II. HANDLING NONDROSOPHILID INSECTS

In general, the techniques that have proven so successful for *D. melanogaster* transformation are used in nondrosophilid insects. Procedures for the handling and microinjection of nondrosophilid insect embryos have been described in detail elsewhere and will not be covered here (O'Brochta and Atkinson, in press; Morris, 1997; Handler, 2000). Researchers considering the use of transformation techniques should visit laboratories that have successfully deployed the procedures. There are many subtleties to the procedures that are better demonstrated by a hands-on experience, and one-to-one communication is often the preferred method to teach the nuances of the different species that have been transformed.

III. THE *P* ELEMENT PARADIGM IS SUCCESSFUL IN NONDROSOPHILIDS, BUT NOT WITH *P* ELEMENTS

The paradigm for insect transformation established by Rubin and Spradling (1982) has proven successful in other insects. This paradigm was developed after the discovery that the P transposable element could be used to integrate exogenous DNA into the germline cells of D. *melanogaster*. Paradoxically, P itself has not been of any use for nondrosophilid insect transformation. Typically, to transform insects, two plasmids are coinjected into the posterior pole of carefully staged preblastoderm embryos. The vector plasmid contains portions of the transposable element, including the inverted terminal repeat (ITR) sequences between which a genetic marker and a gene of interest have been inserted. The second plasmid, the "helper," contains the appropriate gene encoding the transposase that mediates the excision of the transposable element from the vector plasmid, and its subsequent integration into the germline chromosomes. Injected embryos, the G_0 generation, are reared to adulthood and outcrossed, and the resulting progeny, the G_1 generation, are examined for the presence of transgenic individuals. Frequencies of P-mediated transformation of D. *melanogaster* are consistently on the order of 10–30%, representing the percentage of fertile, single G_0 founder matings that yield transgenic progeny. As described later, this range of frequencies also is seen when other transposable elements, such as Hermes and piggyBac, are used as gene transformation vectors in this species (O'Brochta et al., 1996; Handler and Harrell, 1999). However, despite their unequivocal success in D. *melanogaster* and some other drosophilid species (Brennan et al., 1984; Scavarda and Hartl, 1984), P elements have failed as gene vectors in other insect species. For example, early efforts to use P to transform the yellow fever mosquito, Aedes aegypti (Morris et al., 1989), Aedes triseriatus (McGrane et al., 1988), and the African malaria vector, Anopheles gambiae (Miller et al., 1987), resulted in low frequency, apparently nonhomologous integration of the transposable element.

Two possible reasons have been suggested for the failure of P to transform other insects. The first reason concerns the distribution of P elements and the regulation of their transposition. Many transposable elements, including P, have mechanisms that repress their own transposition (Misra and Rio, 1990). These mechanisms are thought to have evolved by selective pressures arising from the mutagenic properties on the host caused by the movement of transposable elements. Therefore, P may be inactive because of transposition repression mediated by endogenous P or P-like elements in the recipient species. However, there is strong evidence that P can be introduced and spread through an insect species. It is widely accepted that P entered the D. *melanogaster* genome at some point over the past 50 years and most likely came from the related species, D. *willistoni* (Engels, 1989; Daniels et al., 1990). The subsequent dispersal of P through D. *melanogaster* has been extremely successful, with the majority of wild *D. melanogaster* populations now containing this element. This spread is remarkable in that it initially carried with it a high negative genetic load that manifested itself as a complex phenotype termed hybrid dysgenesis (Kidwell *et al.*, 1977).

   P elements and *P* element-like sequences also exist in other insect species. Potentially active *P* elements also are present in *D. willistoni* (Daniels *et al.*, 1990), *D. bifasciata* (Hagemann *et al.*, 1992), *D. imaii* (Haring *et al.*, 1995), *D. paulistorum* (Daniels *et al.*, 1984) and *Scaptomyza pallida* (Simonelig and Anxolabehere, 1991), while inactive *P* element sequences have been identified in *D. nebulosa* (Lansman *et al.*, 1987), *D. subobscura* (Paricio *et al.*, 1991), *D. madeirensis* (Paricio *et al.*, 1996) and *D. gaunche* (Miller *et al.*, 1992). Two types of inactive *P* element were identified in the Australian sheep blowfly, *Lucilia cuprina* (Perkins and Howells, 1992). More recently, Lee *et al.* (1999) isolated and characterized a *P* element sequence from the housefly, *Musca domestica*, and used Southern blot analyses to show that several copies of this element were present in the genome. Portions of the housefly *P* element were more similar in DNA and amino acid sequence, and in transcriptional organization to the *P* element-like sequence from *L. cuprina* than they were to the *Drosophila P* element.

   An insight into the evolution of *P* element transposition and repression was obtained when the primary DNA sequences of the genes encoding the *P* transposases from drosophilid species were compared. Witherspoon (1999) examined the rates of synonymous and nonsynonymous substitutions in the four exons of *P* element transposase sequence in 16 *P*-like transposase genes taken from 11 drosophilid species. The first three exons of the transposase gene are involved in both transposition and repression because the *P* element repressor protein is encoded by only these exons. The fourth exon of the *P* element transposase gene is involved only in the transposition function of the transposase. From his analysis of these sequences, Witherspoon (1999) concluded that, over time, selection has acted to conserve the function of both the repressor component of the transposase (exons 0,1,2) and the transpositionally active component of the transposase (exon 3). This finding indicates that, once established in a new host, selection on the host to inactivate *P* element transposition leads to the conservation of repression function. If one assumes that the final exon in the *P* element transposase gene is involved only in transposition, a counter selection then preserves this function so that the *P* element can retain the ability to transfer horizontally to a new species when the opportunity presents.

   Clearly, *P* elements may be more widespread in insects than previously thought. The presence of four copies of this element in the *M. domestica* genome suggests that it once was an active transposable element. The presence of *P*-like elements in insects other than *D. melanogaster* may indicate that there is also an endogenous active system of *P* mobility repression to inactivate these elements once they have invaded a genome. Therefore, this endogenous repression may have, in part, frustrated efforts to use P as a gene vector in nondrosophilid insect species.

The second reason offered for the failure of P in nondrosophilids concerns its requirement for host-encoded proteins and the unique nature of the structure of its cleaved termini during the transposition process. The best mechanistic models of P transposase activity indicate that an oligomer (most likely a dimer) of transposase recognizes single-stranded and duplex DNA binding sites within and adjacent to the ITRs where they then can also interact with the recipient or target DNA (Beall and Rio, 1998). A series of endonuclease reactions followed by strand transfer mediates movement of P from the donor to recipient DNA (Beall and Rio, 1997). P transposes by a cut-and-paste (conservative) mechanism that, in the presence of homologous sequences, involves precise excision and integration of the ITR sequences and any intervening DNA (Figure 2.2; see color insert). In the absence of homologous sequences, excision of P is usually imprecise, most likely arising from the inefficient repair of the long, single-stranded P element ends present during transposition (Beall and Rio, 1997; O'Brochta et al., 1991). A more complex mode of transposition, replicative, has been observed for other transposable elements, such as Tn7, which can also undergo cut-and-paste transposition (May and Craig, 1996). While some integrations of P, such as the tandem array of elements and plasmid sequence seen in the R310.1 transgenic line of D. melanogaster (Rubin and Spradling, 1983), are suggestive of a replicative mechanism of transposition, this mode of transposition has yet to be clearly demonstrated.

P encodes a transposase that has a weak binding affinity for DNA and only a marginally stronger binding affinity for the two transposase binding sites near the termini of the element (Kaufman et al., 1989). These binding sites are 10 base pairs (bp) in length and are located internal to the P element 31 bp ITRs, approximately 40 and 52 bp in distance from the two cleavage sites located in the ITRs (Mullins et al., 1989). The transposase also binds within the 31 bp ITRs, although the interaction with these was only revealed when in vitro assays designed to examine the interaction between transposase and the P element termini during actual transposition were conducted (Beall and Rio, 1998). The pattern of cleavage mediated by the P transposase is unique when compared with that observed for other eukaryotic transposable elements. As for other eukaryote Class II elements, the 3'-end cleavage sites are at the terminal nucleotide of the P element; however, the 5'-end cleavage sites are located 17 bp in from the terminal nucleotides, immediately adjacent to the binding sites of another factor, inverted repeat binding protein (IRBP) (Beall and Rio, 1998). Although other transposable elements also have staggered cuts at their terminal sequences, the longer length of the 17 bp 3'-end overhang is unique to P. IRBP has been proposed to interact with a RECQ-type helicase encoded by the

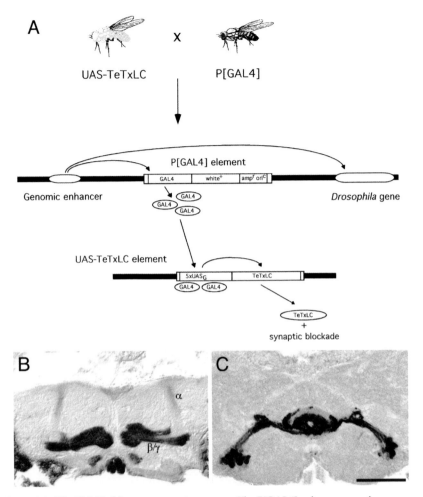

Figure 1.1. The P[GAL4] binary expression system. The P[GAL4] enhancer-trap element contains a minimal promoter region upstream of the GAL4 coding region, insufficient
in itself to induce transcription. After transposition, insertion in a region under the
influence of a local genomic enhance causes the GAL4 gene to be transcribed in a
pattern reflecting the expression of the endogenous gene. The P[GAL4] element can
be used to drive expression of any gene that is placed under the control of the activation sequence, UAS$_G$, in this case the gene encoding the tetanus toxin light chain
(TeTxLC). The P[GAL4] and UAS$_G$-TeTxLC elements are generated separately in
two fly stocks. Crossing these two stocks together brings the two elements together in the progeny, causing ectopic expression of TeTxLC in a defined pattern.
TeTxLC blocks evoked synaptic release of neurotransmitter; thus a "genetic lobot-
omy" is produced in flies of such a cross (Brand and Perrimon, 1993; Sweeney *et
al.*, 1995). (B) Cryostat frontal sections (10 μm) of TeTxLC immunostaining in
adult brains of *Drosophila melanogaster* where light-chain expression is driven by
different P[GAL4] enhancer trap elements. P[GAL4]201Y drives expression of
TeTxLC in the mushroom bodies, principally in the γ-lobe. A faint staining is also
observed in the α and β lobes. (Reprinted with permission from Martin *et al.*,
1998). (C) P[GAL4]C232 drives expression of TeTxLC in the ellipsoid body. The
staining is restricted to a subset of ring neurons, which arborize in two distinct
regions: an inner part surrounding the ellipsoid-body canal and an outer part at the
periphery of the ellipsoid body. The cell bodies on both sides are located in a ros-
tro-lateral-ventral position by reference to the ellipsoid body visible because of the
slightly oblique section. J. R. Martin, unpublished result. Scale bar = 50 μm.

Replicative

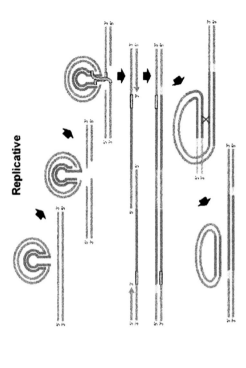

Cut and Paste

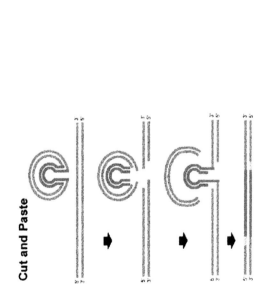

Figure 2.2. Common modes of Class II transposable element mobility. Cut-and-paste refers to a conservative mechanism in which the transposable element (blue lines) excises via a double-strand cut in the donor DNA (gray circle) and inserts into a double-strand cut in the target DNA (gray lines). Following insertion and because of the staggered cut made in the target DNA, a target site duplication (yellow) is created. Note that the copy number of the element does not increase. Replicative transposition initiates with a single-strand cut in the donor DNA and a double-strand cut in the target DNA. Contact is made by the transposable element (light blue lines) with the target DNA, which is subsequently used as a template to synthesize double strands of the element. The integrated donor DNA and one copy of the element are excised by recombination (red ×). This leaves an intact donor DNA molecule and integrated copy of the transposable element flanked by target site repeats (yellow). This type of transgenesis leads to an increase in copy number of the transposable element. 5′ and 3′ refer to the orientation of the DNA strands.

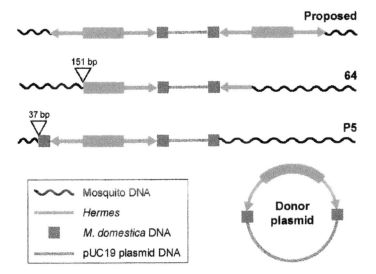

Figure 2.3. Putative model to explain the structure of *Hermes*-mediated integrations into the chromosomes of *Aedes aegypti*. The proposed structure of a replicative transposition intermediate is drawn at top. It consists of two copies of the DNA including the ITRs and intervening DNA flanking the cDNA from the donor plasmid. Below are schematics of the structure of two integrations, 64 and P5, into the mosquito DNA. The numbered arrowheads refer to the size and location of deletions that disrupt the contiguity of donor DNA as it exists in the donor plasmid.

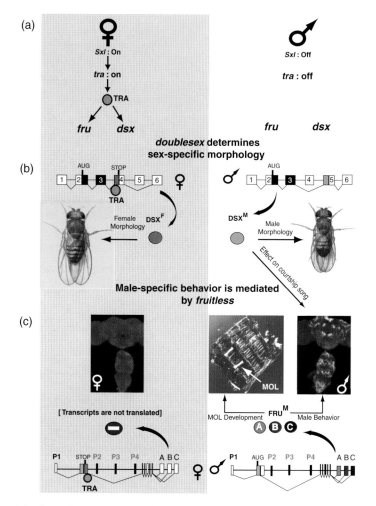

Figure 3.2. Genes determining sexual differentiation in *Drosophila melanogaster* (a) In females the expression of the *Sex-lethal* (*Sxl*) gene controls the production of functional TRA protein by sex-specific splicing. In males *Sxl* is silent and *transformer* (*tra*) is default spliced to make a non-functional protein. (b) TRA controls the sex-specific splicing of *doublesex* (*dsx*) RNA; this gene determines most aspects of sex-specific development (see text). In males *dsx* pre-mRNA is default-spliced to generate the male-specific protein DSX^M that directs male development. In females, TRA protein binds to the *dsx* pre-mRNA, contributing to the production of an alternative splice-form that generates the female-specific DSX^F protein. Note that the two proteins have the exons 1, 2, 3, and 6 in common. However, females retain exon 4, whereas males retain exon 5. For details about the gene products shown, see text; and for an extensive review, see Cline and Meyer (1996). (c) TRA also controls the sex-specific splicing of *fruitless* (*fru*), a gene that influences many features of male-specific behavior. In males, *fru* pre-mRNA from the P1 promoter is default-spliced to produce three alternative full-length proteins (FRU^M), each of which has exon A or B or C in a relatively 3′-region of the mature mRNA. These P1 transcripts are expressed in the male CNS to direct male-specific behavior and formation of the muscle of Lawrence (MOL). In females TRA protein binds to the *fru* pre-mRNA to produce transcripts that have a premature stop-codon, so no functional FRU protein is produced. Note that transcripts (not shown) from promoters P2, P3, and P4 are found in both sexes and appear to be involved in *fru*'s vital function (Goodwin *et al.*, 2000; Anand *et al.*, 2001).

D. melanogaster mus309 locus (Kusano *et al.*, 2001). This gene has been shown to be involved in at least *P* excision (Beall and Rio, 1996). *Mus309* mutants display a decrease in excision frequency and an alteration in the structure of the excision sites following *P* excision (Beall and Rio, 1996). The precise role of IRBP in *P* transposition remains unclear. It has been proposed to participate in the repair of empty sites following excision of *P* (Beall and Rio, 1998). This conclusion is based on the amino acid similarity of IRBP to the Ku protein of the mammalian V(D)J recombination system, which is known, together with DNA-dependent protein kinase proteins, to bind to single-stranded DNA and lead to repair of gaps arising from recombination (Weaver, 1995). Similar to Ku, IRBP also binds the single-stranded 17 bp 3′-end overhangs that arise during *P* element transposition. The involvement of IRBP in *P* mobility demonstrates that host factors are involved in at least *P* excision in *D. melanogaster*. Other insect species may not possess genes that encode IRBP homologs, or they may have paralogous or orthologous IRBP genes that encode proteins in which the tertiary structure is sufficiently different to improperly interact with *P*. Furthermore, the unusually long length of the single-stranded 3′-end overhang generated during *P* element excision in *Drosophila* either may not occur when *P* is placed in nondrosophilid cells, or may be an unsuitable substrate for subsequent transposition in these species.

While the recent progress in successfully developing other transposable elements greatly diminishes, or even eliminates, the need to develop *P* as a generic insect gene vector, the issue of *P* mobility in nondrosophilid species remains. How can it be so successful in dispersing through *D. melanogaster* populations over the past 50 years, but be so ineffective as a gene vector in every nondrosophilid insect in which it has been tried? What do the apparent remnants of *P* elements in some insect species tell us about the origin of *P* and perhaps its ancient distribution among insects? These are fascinating questions, and the answers are perhaps relevant to current endeavors to use other transposable elements as gene vectors for spreading genes through wild insect populations. While *P* as a gene vector in nondrosophilid insects has faded from our attention, there is still much to learn from its evolutionary history and current behavior.

IV. TRANSFORMATION OF NONDROSOPHILID INSECTS BY TRANSPOSABLE ELEMENTS

Four transposable elements, each from a separate transposable element family, can now be used to transform nondrosophilid insects (Table 2.1 and Figure 2.1). The development of these vectors has led to a rush of experiments in which genes and promoters have been introduced into nonhost species.

A. *Hermes* elements

1. Description

Hermes is a member of the *hAT* family of transposable elements. *HAT* elements have been discovered in plants, fungi, insects, nematodes, fish, and humans (Grappin *et al.*, 1996; Huttley *et al.*, 1995; Tsay *et al.*, 1993; Gomez-Gomez *et al.*, 1999; Kempken and Kuck, 1996; Streck *et al.*, 1986; Calvi *et al.*, 1991; Warren *et al.*, 1994, 1995; Coates *et al.*, 1996; Handler and Gomez, 1997; Pinkerton *et al.*, 1999; Bigot *et al.*, 1996; Okuda *et al.*, 1998; Hori *et al.*, 1998; Smit, 1999; Esposito *et al.*, 1999). They are defined by their common overall size, typically 3–5 kb; the conserved sequences of their ITRs; the conservation of at least one, but as many as three, regions within their single open reading frame transposase genes; the generation of 8-bp target site duplications at the site of insertion into DNA; and the type of footprint they leave in the host DNA following excision. The complete sequences of five insect *hAT* elements have been determined or reconstructed. These are the *hobo* from *D. melanogaster* (Streck *et al.*, 1986; Calvi *et al.*, 1991), *Hermes* from *M. domestica* (Warren *et al.*, 1994), *Homer* from *Bactrocera tryoni* (Pinkerton *et al.*, 1999), *hermit* from *L. cuprina* (Coates *et al.*, 1996), and *Hopper* from *B. dorsalis* (Handler and Gomez, 1997). Partially characterized elements include *Hector* from *M. vetustissima* (Warren *et al.*, 1995) and *Huni* from *Anopheles gambiae* (F. H. Collins, D. A. O'Brochta, and P. W. Atkinson, unpublished data). A second *hAT* element sequence from *An. gambiae* has also been recovered as part of the *An. gambiae* genome project (Roth *et al.*, 2000). Portions of what look like *hAT* elements have also been found in *Aedes aegypti* (Stark and James, 1998).

All of these species are within the order Diptera; however, based on their distribution across kingdoms, it is likely that *hAT* elements are spread throughout all groups of insects and other arthropods as well. Indeed, the best characterized *hAT* element exists in plants. The *Activator* (*Ac*) element of maize introduced biological science to the existence of transposable elements and eventually led to the acceptance that genomes were dynamic and could undergo rearrangements independent of sexual recombination mechanisms (McClintock, 1950). A significant observation for this family of elements was that the plant *hAT* elements are mobile in plant species other than their original hosts (for example, Hehl and Baker, 1990). This fueled the expectation that a broad host range would also be a characteristic of insect *hAT* elements, and this hope has been met primarily with the *Hermes* element.

Hermes was isolated from *M. domestica* following demonstration that the related *hobo* element could excise in housefly embryos in the absence of *hobo* transposase (Atkinson *et al.*, 1993). An endogenous active *hAT* element was proposed to be the most likely source of this excision activity. Areas of conservation between three *hAT* elements, *hobo*, *Ac*, and *Tam3*, were used to design degenerate oligonucleotide primers that then were used to amplify internal regions of *Hermes* from the housefly genome (Atkinson *et al.*, 1993; Warren *et al.*, 1994). Inverse gene amplification procedures enabled the entire *Hermes* element to be cloned and reconstructed. *Hermes* is 2749 bp in length, contains 17 bp ITRs, and encodes a transposase protein with a calculated molecular mass of 70 kilodaltons (kDa). The *hobo* and *Hermes* transposases are 55% identical and 70% similar at the amino acid level. *hobo* and *Hermes* ITRs are identical for 10 (left-hand ITR) and 11 (right-hand ITR) out of 12 nucleotides and can cross-mobilize each other, as measured by plasmid-based and chromosomal-based excision assays performed in *D. melanogaster* (Sundararajan *et al.*, 1999). However, there is no direct reciprocity in their ability to cross-mobilize one another: *hobo* elements can mobilize *Hermes* elements at a higher frequency than *Hermes* can mobilize *hobo* (Sundararajan *et al.*, 1999).

Hermes elements have been found in housefly populations from throughout the world (Cathcart *et al.*, submitted). Fourteen populations of flies from four continents were examined and found to have full-length *Hermes* elements present in all of them. Deleted forms of *Hermes* also were present in all of the populations. This distribution was unlike the distribution of *P* and *hobo* in natural populations of *D. melanogaster*, in which large numbers of internally deleted elements are present; and very few, if any, full-length autonomous elements exist. To date, no housefly population devoid of *Hermes* elements has been found.

2. Use as a gene vector

Hermes has been used to transform six species of insects. *Hermes*-mediated transformation of *D. melanogaster* occurs routinely at frequencies of 20–40% (O'Brochta *et al.*, 1996; Pinkerton *et al.*, 2000). Thus, *Hermes* is as efficient as *P* in producing *D. melanogaster* transformants. *Hermes* has been used to generate transgenic *Aedes aegypti* (4% average transformation frequency) (Jasinskiene *et al.*, 1998; Pinkerton *et al.*, 2000; Kokoza *et al.*, 2000; Moreira *et al.*, 2000), *C. capitata* (1%) (Michel *et al.*, 2001), *Stomoxys calcitrans* (4%) (Lehane *et al.*, 2000), *Tribolium castaneum* (1%) (Berghammer *et al.*, 1999), and *Culex quinquefasciatus* (12%) (Allen *et al.*, 2001). The availability of an *Ae. aegypti* white-eyed mutant recipient strain, *kynurenine hydroxylase-white* (kh^w) (Cornel *et al.*, 1997; Jasinskiene *et al.*, 1998), made it possible to use a wild-type copy of the homologous *D. melanogaster cinnabar* gene (cn^+) as a marker for transformation. The enhanced form of the green fluorescent protein (EGFP) gene also was used as the genetic marker in this mosquito and other species (Pinkerton *et al.*, 2000). For *C. capitata*, the cDNA form of the *C. capitata white* gene was used as the genetic marker (Michel *et al.*, 2001). EGFP was the only genetic marker used for detecting transgenics of *S. calcitrans*, *Cx. quinquefasciatus*, and *T. castaneum* (Lehane *et al.*, 2000; Allen *et al.*, 2001 Berghammer *et al.*, 1999).

Two types of integration have been observed for *Hermes* elements: events arising from cut-and-paste transposition of only the transposable element into the genome, and those arising from the insertion of the element and flanking plasmid

DNA into the genome. Hermes-mediated transformation of D. melanogaster, C. capitata, and S. calcitrans results in the integration of only the Hermes element and any additional sequences located within it (O'Brochta et al., 1996; Michel et al., 2001; Lehane et al., 2000). The integrated sequences are delimited by the terminal nucleotides of the Hermes element, and 8 bp duplications are created at the insertion site (Streck et al., 1986; Sarkar et al., 1997a,b). The sequence of these conforms to the consensus sequence of target-site duplications created by the transposition of insect hAT elements. The integration of Hermes elements into these three species is therefore expected for cut-and-paste Class II insect transposable elements and was predicted from Hermes interplasmid transposition assays performed in each of these species (O'Brochta et al., 1996; Sarkar et al., 1997a; Lehane et al., 2000).

Hermes-mediated transformation of the mosquitoes, Ae. aegypti and Cx. quinquefasciatus results in integration of both the Hermes element and flanking M. domestica DNA (derived from the original cloning steps) and plasmid sequences (Jasinskiene et al., 1998; Pinkerton et al., 2000; Allen et al., 2001). In Ae. aegypti, these events require Hermes transposase, because transformation does not occur in the absence of a coinjected helper plasmid containing the Hermes transposase gene (Jasinskiene et al., 2000). Equivalent experiments have been performed in Cx. quinquefasciatus and the similarity in both the structure of the integrations and the relatively high frequency (12%) of transformation suggests that these, too, are dependent on the presence of Hermes transposase (Allen et al., 2001). The structures of cn^+-containing Hermes elements in Ae. aegypti were examined by Jasinskiene et al., (2000). Breakpoints were found in plasmid DNA, flanking M. domestica DNA, and in Hermes itself (Figure 2.3; see color insert). A number of possible mechanisms—including aberrant resolution of cointegrate structures during a replicative transposition event, or transposase-mediated illegitimate recombination—may account for this type of integration. However, at this time an insufficient number of transformed lines exists to reveal which of a number of alternative mechanisms is favored. It is clear from EGFP-containing transgenic lines of both Ae. aegypti and Cx. quinquefascitus that at least two copies of the Hermes element are present, and these flank a complete and intact copy of the pUC plasmid DNA that, with Hermes, constituted the original plasmid vector (Pinkerton et al., 2000; Allen et al., 2001). This arrangement of integrated plasmid plus transposable element appears similar to that reported for the R310.1 transgenic line of D. melanogaster in which P and pUC plasmid sequences were integrated into the genome as a tandem array (Rubin and Spradling, 1983).

Hermes can, however, transpose in a cut-and-paste manner in mosquitoes. Zhao and Eggleston (1998) recovered transformed cell lines of An. gambiae in which only Hermes element DNA, had integrated. The expected 8 bp duplications were present at the target sites. Plasmid-based Hermes transposition assays performed in An. gambiae, Ae. aegypti, and Cx. quinquefascatus led to the recovery of *Hermes* elements that had transposed into target plasmid DNA in the predicted way (Sarkar *et al.*, 1997b; M. L. Allen and P. W. Atkinson, unpublished data; D. A. O'Brochta, F. H. Collins, and P. W. Atkinson, unpublished data). The sequences transposed were delimited by the terminal nucleotides of the transposable element, and an 8 bp duplication was created at the target site. However, the design of these plasmid assays excluded the selection of the types of *Hermes* transposition events recovered from transgenic mosquitoes. Evidence has been obtained for cut-and-paste transpositions of *Hermes* in the chromosomes of *Ae. aegypti*. An autonomous *Hermes* element containing the *Hermes* transposase placed under the control of the *heat shock protein 70* gene (*hsp70*) promoter of *D. melanogaster* was integrated into the *Ae. aegypti* genome using standard procedures. While the initial integration of this element also involved flanking plasmid sequences, subsequent remobilization of the *Hermes* element in somatic nuclei appears to have been achieved by the conventional cut-and-paste mechanism of transposition (R. Hice, C. S. LeVesque, D. A. O'Brochta, and P. W. Atkinson, unpublished data). This supports the plasmid-based *Hermes* assays performed by Sarkar *et al.*, (1997b), which predominately measure transpositional activity in the somatic nuclei of the developing embryo. The molecular basis of this apparent difference in the mode of *Hermes* integration between germline and somatic mosquito nuclei remains unknown.

3. Current issues arising from the use of *Hermes* elements in insects

Hermes has proved an effective gene vector in several insect species. Since both *Hermes* and *hobo* are active elements for which genetic transformation techniques have been established in *Drosophila*, the *Hermes* system is the only one of the four transposable element systems now available for nondrosophilid insect transformation that has been tested for cross-mobilization in the presence of endogenous elements of the same family (Sundararajan *et al.*, 1999). These experiments clearly showed that cross-mobilization can occur. Malacrida and colleagues have obtained evidence proving that genetically transformed strains of *C. capitata* containing *Hermes* are unstable for this element (A. Malacrida, personal communication). The presence of a *hobo*-like element in *C. capitata* has been proposed to be the source of this instability (Handler and Gomez, 1996). These results have served to highlight the possibility of cross-mobility in insects containing endogenous transposable elements that are in the same family as the transposable element being used as a gene vector. The *hAT* element family system has provided the best means by which this issue can be addressed directly through experimentation in insects. The outcomes of these experiments have illustrated that the composition of the target insect species with respect to endogenous transposable elements is a key consideration when choosing a transposable element system. In addition, the two different modes of *Hermes* element integration into mosquito genomes have not been described for any other transposable element, except for the example of P in the R310.1 line of D. melanogaster. This raises a question about the role that host-encoded factors may play in the transposition of transposable elements into insect genomes. As described earlier, host-encoded factors are known to play a role in P mobility. The role that such factors may play in the mobility of Hermes and other transposable elements is a matter of considerable interest and importance, particularly if the mobility and stability of these elements is to be controlled subsequent to initial integration in the target genome.

B. *Mariner* elements

1. Description

Mariner elements are widely distributed among arthropods and, together with the related Tc1 element from *Caenorhabditis elegans*, represent perhaps the most dispersed group of transposable elements in nature (Robertson, 1995). *Mariner* elements are approximately 1.3 kb in length and contain a transposase gene flanked by ITRs that are approximately 30 bp in length. Their transposases consist of three domains, an N-terminal DNA binding domain that contains a bipartite helix–turn–helix motif, a nuclear localization signal domain, and a catalytic domain, which, in *mariner* elements, contains a DDD motif. This motif is thought to be critical in the binding of the divalent cation required as a cofactor by polynucleotide kinase. The three aspartic acid residues form part of the catalytic site but are separated in the transposase primary sequence by more than 90 residues (between the first and second aspartic acid residues) and approximately 34–35 residues (between the second and third aspartic acid residues) (Craig, 1995). Based on similarities in the amino acid sequence of their transposases, the lineage of *mariner* elements can be traced back to several bacterial insertion sequences such as IS3 and IS630 (Plasterk et al., 1999). A comprehensive analysis of the phylogenies of *mariner* elements led to the classification of five subfamilies (*capitata, cecropia, irritans, mauritiana,* and *mellifera*) within insects alone (Robertson and MacLeod, 1993). *Mariner* elements can be present in thousands of copies per genome equivalent and, in many cases, members of more than one subfamily of *mariner* elements may be present in the same genome. The construction of phylogenetic trees based on the nucleic acid and amino acid sequences of *mariner* elements results in several cases in clear divergences from accepted phylogenetic trees that are based either on the sequences of host chromosomal genes or morphological characters (Capy et al., 1994; Maruyama and Hartl, 1991; Lohe et al., 1995; Garcia-Fernandez et al., 1995). It is now accepted widely that these differences arise from the horizontal transfer of *mariner* elements among species. However, as for the P element, direct evidence of *mariner* horizontal transfer has yet to be obtained. The distribution of *mariner* elements in insects has justifiably fueled interest in developing them as robust gene transfer vectors in insects. More recently, this desire has been further enhanced by the demonstration that *mariner* is an effective gene transfer vector in such diverse organisms as zebrafish (Fadool *et al.*, 1998), *Bombyx mori* cells (Wang *et al.*, 2000a), housefly (Yoshiyama *et al.*, 2000), the protozoan parasite *Leishmania* (Gueiros-Filho and Beverley, 1997), mammalian cells (Zhang *et al.*, 1998), and chickens (Sherman *et al.*, 1998).

The abundance of *mariner* elements in insect genomes makes it difficult to isolate the few copies of the element that may encode a functional transposase. To date, only one naturally occurring, functional *mariner* element, Mos1 from *D. mauritiana*, has been isolated from insects (Medhora *et al.*, 1991). A second, *Himar*, is a consensus element based on the sequence of various copies of *mariner* isolated from the horn fly, *Hematobia irritans* (Lampe *et al.*, 1996). *Himar* is mobile in bacteria (Rubin *et al.*, 1999) and in mammalian cells (Zhang *et al.*, 1998), but has yet to prove successful as a gene vector in insects. Indeed, *Himar* displays extremely limited mobility in *Drosophila*, even after it has been introduced into the genome by *P*-mediated transformation (Lampe *et al.*, 2000).

The regulation of element mobility is a major issue that needs to be addressed if *mariner* elements are to be used as robust gene vectors in insects. Based on phylogenetic distributions and the high copy numbers seen in many species, it is clear that *mariner* elements can be successful at invading, and then spreading through, insect species. This is an important observation since one of the applications driving the development of transgenic nondrosophilid technology has been the desire to quickly spread beneficial genes through insect populations. The success of *mariner* elements at spreading through genomes and populations might be expected to lead to mechanisms that eventually regulate the copy number of active *mariner* elements. Lohe and Hartl (1996a) and Hartl *et al.*, (1997) examined the regulation of *mariner* elements and identified at least two mechanisms, overproduction inhibition and dominant-negative complementation, that could regulate the copy number of *mariner* elements. Overproduction inhibition arises from excessive production of transposase and was proposed to decrease the excision frequency of *mariner* elements in the *D. melanogaster* genome, most likely though a posttranscriptional mechanism (Lohe and Hartl, 1996b). Dominant-negative complementation has been proposed to be a consequence of the formation of heterodimers of transposase that contain active and dysfunctional mariner transposase molecules, leading to oligomers and/or heterodimers that are less active than the wild-type form of the enzyme (Lohe and Hartl, 1996b).

2. Use as a gene vector

As mentioned earlier, Mos1 from *D. mauritiana* is the only naturally occurring *mariner* element that is active. Mos1 has been used to transform *D. melanogaster*, *D. virilis*, *M. domestica*, and *Ae. aegypti* (Lidholm *et al.*, 1993; Lohe and Hartl, 1996b; Yoshiyama *et al.*, 2000; Coates *et al.*, 1998). The *mariner* element was also used to transform *Leishmania* (Gueiros-Filho and Beverley, 1997), zebrafish

(Fadool *et al.*, 1998), and chickens (Sherman *et al.*, 1998). Coates *et al.*, (1997) developed interplasmid transposition assays that showed that Mos1 could accurately transpose in at least three species of non-drosophilid insects, Ae. *aegypti*, L. *cuprina*, and B. *tryoni*.

Himar is a synthetic element based on the consensus sequence obtained from elements isolated from H. *irritans* (Lampe *et al.*, 1996). The genetic modification of Himar in order both to understand the molecular basis of its movement and to isolate hyperactive forms of this element has received much attention. Lampe *et al.*, (1999) utilized a bacterial colony papillation assay based on the successful assays established for the *Escherichia coli* element, Tn5 (Krebs and Reznikoff, 1998) and isolated two Himar mutants that displayed increased levels of transposition in E. *coli*. This type of strategy will no doubt lead to new forms of Himar, and some most likely will also have hypermobility properties in insects.

As for other members of the *mariner/Tc1* superfamily of elements, Mos1 and Himar insert only at TA dinucleotide sequences and create 2 bp target site duplications. In D. *melanogaster* transformants and in transpositions arising from plasmid mobility assays performed in B. *tryoni*, L. *cuprina*, and Ae. *aegypti*, the *mariner* sequences that are integrated are delimited by the terminal nucleotides of the element ITRs and resemble the products of cut-and-paste transposition (Lidholm *et al.*, 1993; Coates *et al.*, 1996, 1999). In Ae. *aegypti* transgenics, most transformed lines contain Mos1 elements that have integrated in the same manner; however, some of the lines contained elements together with flanking plasmid DNA sequences, an event similar to what was seen with Hermes integrations in this species (Coates *et al.*, 1998, 2000). When Mos1 transposase protein was used instead of helper plasmid, the frequency of these non-cut-and-paste integration events increased (Coates *et al.*, 2000). Remarkably, Mos1 was able to catalyze the integration of ~12 kb of exogenous DNA in Ae. *aegypti* (Coates *et al.*, 1998).

3. Current issues arising from the use of *mariner* as a gene vector in insects

Mariner elements are enigmatic. Their wide distribution among arthropods, their high copy number in individual species, and the compelling evidence of their history of horizontal transfer all suggest that they should be versatile gene vectors in insects. Yet, other than for Ae. *aegypti*, they are not. Even in the case of Mos1 in D. *melanogaster*, transformation frequencies are low, and there is no example of it being subsequently remobilized in the germlines of transgenic lines despite the introduction of Mos1 transposase (Lidholm *et al.*, 1993). As described previously, the abundance of *mariner* elements in individual species has made it difficult to isolate active forms of this element, so that only two active elements have been isolated and tested. This has probably hampered the utilization of *mariner* elements as gene vectors in insects. Yet the issue of the regulation of these elements remains to be determined. Are they simply so active as natural gene vectors that insects have been selected to express efficient repression mechanisms to deal with their invasion? Or have these elements been selected for their own repression mechanisms that ensure their propagation through generations? Can *mariner* be used as vectors in those species that contain thousands of copies of elements (only a few of which may encode active transposase)? The isolation of a *mariner* element that is a robust gene vector in many nondrosophilid species will enable these questions to be addressed.

C. *PiggyBac* elements

1. Description

The *piggyBac* element was isolated from lepidopteran cell cultures on the basis of its mobility, and as a consequence it has been subsequently developed into an efficient gene vector in insects. *PiggyBac* was identified as an insertion sequence that caused a plaque morphology mutation in a *Galleria melonella* nucleopolyhedrosis virus that was being passed through a cell line of the cabbage looper, *Trichoplusia ni* (Fraser *et al.*, 1983; Cary *et al.*, 1989). *PiggyBac* is 2.5 kb in size and has 13 bp ITRs (Elick *et al.*, 1995). The *piggyBac* element inserts at TTAA sequences in the genome and, upon insertion, generates a duplication of this sequence (Cary *et al.*, 1989). Unlike any other insect transposable element so far characterized, *piggyBac* excises with absolute precision, so that no evidence of *piggyBac* remains at the donor site following excision (Fraser *et al.*, 1996). *PiggyBac* comprises its own family of transposable elements, because there are no other transposable elements identified so far that have similar structural or behavioral characteristics. No specific host factors required for *piggyBac* mobility have yet been identified; however, the *piggyBac* ITRs do interact with proteins present in cell nuclear extracts prepared from *Plodia* and *Tricophlusia* cell lines (Bauser *et al.*, 1999). The identity of these proteins and whether they are absolutely necessary for *piggyBac* transposition are unknown.

2. Use as a gene vector

PiggyBac has been used to transform a wide range of insect species, including *D. melanogaster* (Handler and Harrell, 1999), *C. capitata* (Handler *et al.*, 1998), *B. dorsalis* (Handler and McCombs, 2000), *Anastrepha suspensa* (Handler and Harrell, 2001), *M. domestica* (Hediger *et al.*, 2001), *Tribolium castaneum* (Berghammer *et al.*, 1999), *Ae. aegypti* (Kokoza *et al.*, 2001), *An. albimanus* (O. P. Perera, R. Harrell, and A. M. Handler, personal communication), *An. stephensi* (M. Jacobs-Lorena, personal communication), *An. gambiae* (Grossman *et al.*, in press), *Bombyx mori* (Tamura *et al.*, 2000), and *Pectinophora gossypiella*

(Peloquin *et al.*, 2000). In all cases genomic integration has been by apparent cut-and-paste transposition of the *piggyBac* element. Transformation frequencies range from 1% to 60%, but these estimations should be treated with caution since pair matings are not possible in many species. In these cases, injected G_0 individuals are mated in pools, making it impossible to trace the lineage of transgenic G_1 insects to a particular G_0 individual. Nonetheless, *piggyBac* is proving to be a very successful gene vector in insects and, of the four elements described in detail here, is perhaps proving the most robust at this time. The native *piggyBac* transposase promoter has been found to drive the expression of sufficient amounts of transposase to mediate transposition of *piggyBac*, although the *hsp70* promoter is now more widely used in helper plasmids, and it appears to increase the transformation efficiency (Handler and Harrell, 1999). The many transformed lines generated are stable in the absence of *piggyBac* transposase. To date, no published information regarding the subsequent remobilization of *piggyBac* has been reported. However, based on the mobility properties of many other insect Class II elements, it seems likely that *piggyBac* could be remobilized following introduction into a new genome. Initial reports using an enhancer-trapping technique identical in principle to that developed for *P* indicate that, at least in this species, *piggyBac* is as efficient as *P* in generating new insertions into the *D. melanogaster* genome (E. A. Wimmer, personal communication).

3. Current issues arising from the use of the use of *piggyBac* as a gene vector in insects

PiggyBac is proving to be a highly mobile transposable element vector and appears to have an almost unlimited host range in insects. The experience of many research groups with *piggyBac* has been one of fairly quick success in generating transgenics once appropriate vectors have been constructed and injections commenced. This is not surprising given that the *piggyBac* element was originally recovered through its ability to jump from (presumably) the chromosomes of *T. ni* into a baculovirus. Originally, because of its initial detection in viruses infecting *T. ni* cells, *piggyBac* was thought to have originated in *T. ni*. However, recent data have challenged initial thoughts on the origin and distribution of *piggyBac* (Handler and McCombs, 2000). The discovery of *piggyBac* DNA sequences in the genome of the oriental fruit fly, *B. dorsalis*, that are more than 95% identical to the *T. ni* element suggests that *piggyBac* may be widely distributed in insects and that, unlike *mariner* and *hAT* elements, which can show significant sequence variation both within and between species, the *piggyBac* primary sequence may be strongly conserved (Handler and McCombs, 2000). The abilities of *piggyBac* to transpose from the genome to a virus and to genetically transform insects as diverse as those belonging to the lepidoptera, coleoptera, and diptera are perhaps consistent with a promiscuity that may lead to a wide distribution across insect species. The question of whether the *piggyBac* element currently deployed as a gene vector will interact with, and be cross-mobilized by, these endogenous elements remains unaddressed. This will be of key importance in determining whether this element will be stable in those species. The outcomes of these studies, when added to those in progress on the interactions between *hAT* elements, will provide guidance concerning the use of Class II transposable elements as gene vectors in insects.

D. *Minos* elements

1. Description

The *Minos* element belongs to the Tc1 family of transposable elements. *Minos* was isolated serendipitously through the sequencing of ribosomal DNA from *D. hydei* (Franz and Savakis, 1991). It is 1.4 kb in size and, unlike *hAT* elements, *mariner* elements, and *piggyBac*, *Minos* has long ITRs of 254 bp in length. Furthermore, unlike the other elements, the *Minos* transposase gene contains an intron (Franz and Savakis, 1991). The DDE motif that defines, in part, members of the *Tc1* family of transposable elements is present in the *Minos* transposase and differs in sequence from the analogous motif, DDD, in the *mariner* family of elements. *Minos*, like *mariner*, inserts only at TA dinucleotide sites and, upon doing so, creates target-site duplications (Loukeris *et al.*, 1995a,b).

2. Use as a gene vector

Minos has been used to transform *D. melanogaster* (Loukeris *et al.*, 1995b), *C. capitata* (Loukeris *et al.*, 1995a), and the mosquito *An. stephensi* (Catteruccia *et al.*, 2000). Transformation frequencies are on the order of 5%. In the majority of cases, transposition into the target genome is by cut-and-paste transposition with the creation of TA target-site duplications at the site of insertion. *Minos* has also been shown to be capable of transposition into lepidopteran (*Spodoptera frugiperda*) and mosquito (*Ae. aegypti* and *An. gambiae*) cells maintained in culture, indicating that it may also prove to be useful as a gene vector in these species (Klinakis *et al.*, 2000; Catterccia *et al.*, 2000).

3. Current issues arising from the use of *Minos* as a gene vector in insects

Minos most likely will have a broad host range if it proves to be as versatile a gene vector as the *Tc1* elements of nematodes. A further issue is the possible degree of interaction between *Minos* elements and *mariner* elements in insects. Both are members of the *Tc1–mariner* superfamily and so share many structural and functional characteristics. Since only three species of insect have so far been transformed with Minos, this remains unknown. Lampe *et al.* (2000) examined whether purified Himar transposase could recognize and bind to the ITRs of mariner elements obtained from other insect species and found at best weak interactions. The functional significance of this, however, remains unknown. Using a functional assay for Tc element movement, van Leunen *et al.* (1993) found no evidence for cross-mobility of the nematode elements Tc2 and Tc3. Although far from definitive, these data indicate that there may be minimal levels of interactions between members of different subfamilies of the mariner/Tc1 superfamily.

V. INSECT TRANSFORMATION AND GENETIC ANALYSIS

It seems likely that the four transposable elements now used to genetically transform nondrosophilid insects will prove to have reasonably similar performance characteristics in terms of host range and transformation frequencies. As discussed previously, the precise properties of each element in the new host species will need to be determined so that, as much as possible, factors such as cross-mobilization and instability caused by the presence of endogenous elements of the same family can be anticipated and overcome. As such, this presents a very different situation than the one that confronts Drosophila geneticists working with P or hobo in strains devoid of these elements.

Anticipating the behavior of any transposable element in new species can be addressed in part by understanding the molecular basis of transposition of each of these elements. Collectively we know very little about how these elements are regulated and what, if any, host factors are required for their movement. These studies will be imperative if the elements are to become part of a larger strategy aimed at releasing genetically engineered insects into the field in order to combat medically or agriculturally important pest species. These current shortcomings notwithstanding, we are now in a position to use these elements as genetic tools in several insect species. A detailed understanding of the molecular basis of P transposition was not required for its use as a genetic tool in D. melanogaster and, likewise, the four elements described here can now in principle be used in gene tagging and enhancer trap screens in nondrosophild insects. For example, if we take the case of one mosquito species, Ae. aegypti, we now have in place the following:

1. At least three transposable element systems, Hermes, Mos1, and piggyBac, which can be used to transform it
2. One conventional genetic marker (cn^+) and several fluorescent markers (with nonoverlapping spectral properties) that can be used as genetic markers
3. Promoters that are known to produce tissue-specific expression of marker genes in transgenic lines

These features can be used to generate the types of strains now common in *D. melanogaster* that are used to explore basic and applied science problems. For example, the transposase of one type of element can be introduced into a species using a second type of element to create a "jump-starter" strain. This strain could be marked with one type of fluorescent protein gene. A second "ammunition" strain can be developed using those elements responsive to the transposase in the jump-starter strain. This would be marked with a second fluorescent protein gene that has nonoverlapping spectral properties from that used in the jump-starter strain (see Horn and Wimmer, 2000, for a list of the various forms of fluorescent protein genes used in insects and their spectral properties). The strains can be crossed and progeny containing both markers examined for desired phenotypes. The jump-starter transposase can then be removed from these lines by outcrossing. These techniques are not new to genetic analyses, but can now be applied to insects of medical and agricultural importance to identify genes that play critical roles in maintaining the vector or pest status of a given species. Combined with the introduction of genomic approaches to several of these species, insect scientists will now have at their disposal many of the tools that up until now have been unavailable.

VI. PROMOTER ANALYSIS IN TRANSGENIC INSECTS

The use of promoters in transgenic insects can be classified into two areas that are not necessarily mutually exclusive. One concerns the use of promoters to drive the expression of marker genes. This has not been a significant issue in *D. melanogaster* genetics where a wealth of marker genes (each with their own promoter) and complementary mutants strains exist. The second area concerns the more traditional analysis of promoter function in which the expression of a reporter gene placed under the control of a putative promoter is monitored to gain insights into which promoter sequences are required for function.

A. Promoters used to drive the expression of marker genes in nondrosophilid insects

At least eight *D. melanogaster* promoters have been shown to function in heterologous insects. Possibly the most widely used is the *hsp70* promoter. Its expression is not always strictly regulated by temperature in heterologous organisms, and there is evidence of constitutive activity in nonhomologous experiments (Morris *et al.*, 1991). A number of *D. melanogaster* general cell promoters—including those from the *polyubiquitin, metallothionine,* and *actin5c* genes—as well as various viral promoters have been shown to work in heterologous species (Handler *et al.*, 1998; Olson *et al.*, 1996; Matsubara *et al.*, 1996). Perhaps one of the most unexpected and fortuitous discoveries in the development of transformation systems has been the correct functioning of the *D. melanogaster* cn^+ gene in the corresponding mutant strain of *Ae. aegypti* (Cornel *et al.*, 1998). The cn^+ gene encodes the enzyme kynurenine hydroxylase, which is part of the ommochrome biosynthetic pathway in both *D. melangaster* and *Ae. aegypti*. Jasinskiene *et al.* (1998) used a 4.7-kb DNA fragment containing a genomic copy of cn^+ (which included 2.7 kb of promoter sequence and an intron within the cn^+ gene) to transform this mosquito. The use of this marker enabled transgenics generated by the *Hermes* and *Mos*1 element to be easily identified (Jasinskiene *et al.*, 1998; Coates *et al.*, 1998), and it was a major component of the development of a successful transformation system for this species. Clearly, the *D. melanogaster* cn^+ gene is sufficiently similar to the *Ae. aegypti* homolog in that it can function in this species. Moreover, the *D. melanogaster* cn^+ gene promoter is recognized by the transcription machinery in *Ae. aegypti* required for its proper stage- and tissue-specific expression and, subsequently, the intron is correctly spliced out. If this level of conservation of function is a general phenomenon, then this bodes well for the expression of *D. melanogaster* genes in nondrosophilid insects.

The successful development of nondrosophilid transformation in insect species where marker gene/recipient strain combinations do not exist is dependent on the use of fluorescent protein genes, such as EGFP, as genetic markers. What has been important in insect species in which this marker has been used has been the restriction of these proteins either to specific organelles or to terminally differentiated tissues. The *D. melanogaster* *actin5C* promoter confines fluorescent proteins to the cytoskeleton and has been used to identify transgenic *Ae. aegypti* (Pinkerton *et al.*, 2000), *D. melanogaster* (Pinkerton *et al.*, 2000), and stable fly, *S. calcitrans* (Lehane *et al.*, 2000). The *D. melanogaster* *polyubiquitin* gene promoter has been used to identify transgenics in a range of insect species as diverse as mosquitoes and fruit flies (Handler *et al.*, 1998; Handler and Harrell, 1999; A. M. Handler, personal communication); however, there have been some recent concerns that high levels of expression of fluorescent proteins from this promoter may lead to lethality in some *Ae. aegypti* strains (N. Jasinskiene and A. A. James, personal communication).

Using three *Pax-6* homodimer binding sites derived from *D. melanogaster*, Wimmer and colleagues constructed a synthetic promoter (Berghammer *et al.*, 1999; Horn and Wimmer, 2000). *Pax* genes encode a family of transcription factors, with *Pax-6* producing a factor specifically involved in eye and central nervous system development (Callaerts *et al.*, 1997). The binding sites for the *Pax-6* transcription factor are small, typically on the order of 15 bp in length (Czerny and Busslinger, 1995). This promoter has proven successful in generating detectable levels of fluorescent protein in a range of insects (Berghammer *et al.*, 1999; Horn and Wimmer, 2000) but is better detected in strains lacking pigment in the adult eye. Nevertheless, it can be detected in pigmented adult eyes and in the anal pads (or homologous structures such as the larval anal papillae in mosquitoes) and should prove to be as close to a "universal" promoter as possible in insects.

Another *Drosophila* actin gene promoter has been found to function in *Cx. quinquefasciatus* and may therefore be of general use in insects. The *actin88* gene of *D. melanogaster* is expressed specifically in the indirect flight muscles during the pupal and adult life stages. A 1-kb fragment from the *actin88* promoter is sufficient for the tissue- and temporal-specific expression of this gene (Barthmaier and Fyrberg, 1995). Allen *et al.*, used this promoter to direct tissue and time specific expression of green fluorescent protein to the indirect flight muscles of transgenic *Cx. quinquefasciatus*, indicating that, as for the cn^+ promoter in transgenic *Ae. aegypti*, there is sufficient conservation of this promoter to enable its correct function in a heterologous species (M. L. Allen and P. W. Atkinson, unpublished data).

B. Tissue- and stage-specific promoters

One of the most powerful applications of transgenic technology is the functional evaluation of DNA sequences that control the transcription of genes. Standard analyses involve determining those DNA sequences that comprise the core promoter elements and enhancers that direct tissue-, sex-, and stage-specific gene expression. Putative control DNA sequences are linked covalently in cis with a readily assayed reporter gene, and the ability of these constructs to mimic the expression of the endogenous gene from which the promoters were derived is evaluated.

Promoters derived from genes expressed constitutively in all cell types of an animal can be assayed *in vitro*, in cultured cells (Fallon, 1991; Zhao and Eggleston, 1999), or in whole organ culture (Morris *et al.*, 1995). However, most *in vitro* preparations and cell cultures do not provide all of the features of differentiated adult tissues and may lack trans-activating proteins. Furthermore, isolated adult insect organs and tissues rarely undergo cell division, which can often be a prerequisite to new gene expression. Therefore, these procedures rarely are sufficient to evaluate stage-, tissue-, and sex-specific gene expression.

1. Homologous transformation

The relatively recent development of transgenesis in nondrosophilid insects accounts for the low number of published studies of promoter function in homologous systems. A transient assay technique involving microinjection into embryos of plasmids or DNA fragments was used to assay the expression of a reporter gene controlled by the *alcohol dehydrogenase* promoter in larvae of *D. melanogaster* (Martin *et al.*, 1986). Using this approach, Shotkoski *et al.* (1996) were able

to define DNA sequences of the γ-aminobutyric acid receptor gene (*Rdl*) of
Ae. aegypti necessary for expression in the mosquito embryos. However, as was
observed for *D. melanogaster,* this technique did not allow the evaluation of
promoters expressed following the onset of metamorphosis.

Putative promoter fragments of two genes—*Maltase-like I* (*Mal-I*) (James
et al., 1989) and *Apyrase* (*Apy*) (Smartt *et al.,* 1995), specifically expressed in the
adult salivary gland of *Ae. aegypti*—were evaluated following *Hermes*-mediated
stable integration of promoter–reporter constructs into the germline (Coates *et al.,*
1999). Fusions of approximately 1.5 and 1.6 kb of the *MalI* and *Apy* promoters,
respectively, were made with the firefly *luciferase* (*luc*) gene. Evaluation of the
expression patterns in larvae, in the heads, thoraces, and abdomens of adults, and
in the salivary glands from dissected adult females showed that both the *MalI* and
Apy promoters were able to direct the expression of the reporter gene with the
correct tissue-, stage-, and sex-specificity. Expression of the reporter genes was low,
and neither Northern analyses nor immunodetection techniques could be used
to measure the gene activity. However, the assay for the *luc* gene was sufficiently
sensitive so as to reveal the promoter activity in these dissected tissues. This work
was the first example of the assay of nondrosophilid promoters in a homologous,
stable transformation system.

The promoter of the *vitellogenin* (*Vg*) gene of *Ae. aegypti* was recently
assayed in a stable transformation system (Kokoza *et al.,* 2000). This gene encodes
the major yolk protein found in mosquito embryos, and its expression is tightly
regulated following a blood meal. A 2.1 kb putative *Vg* promoter fragment was
cloned at the 5′ end of the β-galactosidase-encoding open reading frame and, in
a separate construct, a mosquito defensin-encoding gene was placed under the
control of this promoter. *Hermes*-mediated integration of these constructs and
the subsequent assay of the transformed lines showed that the *Vg* promoter could
direct expression in adult female fat body with the temporal pattern consistent
with that of the endogenous gene. Unlike what was seen with the salivary gland
gene promoters (Coates *et al.,* 1999), the *Vg* promoter was able to direct abundant
expression of the reporter gene.

Hermes-mediated integrations of the *Ae. aegypti* carboxypeptidase
(*AaCP*) promoter–reporter constructs showed that a 1.4 kb putative promoter
fragment of the *AaCP* gene linked to the *luc* coding sequences expressed *luc* RNA
abundantly in the adult midgut with an expression profile characteristic of the
endogenous gene (Moreira *et al.,* 2000). Furthermore, promoter sequences from
an *An. gambiae* homologous gene, *AgCP*, also demonstrated correct function in
transformed *Ae. aegypti.*

The *Minos*-based stable transformation system (Loukeris *et al.,* 1995a,b)
was used to evaluate promoters from genes encoding male-specific serum
polypeptides (*MSSP*) in the medfly, *C. capitata* (Christophides *et al.,* 2000a,b).
These genes belong to a family of proteins that show structural similarities to

odorant binding proteins. One member of the family, MSSP-α2, is specifically specifically in adult male fat body, whereas another member, MSSP-β2, is expressed in the midgut of both sexes. The researchers emphasize that although the promoters of both genes show a high degree of similarity in their primary structure, they have distinct developmental expression profiles. This differential expression profile was recapitulated in a transgenic C. capitata following transformation of different promoter constructs driving the lacZ gene. These results emphasize the utility of Minos to evaluate promoter function in this important agricultural species.

A surprising result was obtained when Crisanti and his colleagues, using the Minos-based system, introduced into An. stephensi the EGFP gene that had been placed under the control of the trypsin promoter from the Antrpy1 gene of An. gambiae (A. Crisanti, personal communication). Trypsin is produced in response to a blood meal and is one enzyme responsible for the breakdown of proteins in the blood meal. It is only expressed in female mosquitoes, since male mosquitoes do not take blood meals. In transgenic lines of An. stephensi, EGFP expression was found to be confined to the midguts of males, but EGFP was not expressed in females. Although these studies are still incomplete, they provide a sobering reminder of the complexity of the systems we are exploring.

Despite the difficulties of generating and maintaining transgenic lines in these nondrosophilid insects, homologous transformation experiments remain the best test for true promoter function. For this reason we can expect that researchers will look for improvements of the technology that will lessen the labor burden associated with these experiments.

2. Heterologous transformation

The evaluation of promoter function in heterologous species was done for two reasons. First, and most obvious, it was the only way to evaluate promoters derived from species in which transgenesis was not available. The relative ease of transforming D. melanogaster did not require a large investment of effort to design and implement experiments to test heterologous promoter function in this species. These experiments were not based, however, on optimism alone, for it had been known for some time that regulated promoters such as that of the hsp70 gene from D. melanogaster transiently introduced into mammalian cells would respond with increased transcription following an exogenously supplied stimulus (Corces et al., 1981). However, the behavior of sex-, tissue-, and stage-specific insect promoters in heterologous species remained unknown.

The first results were presented by Mitsialis and Kafatos (1985) when they showed that representative promoters of two distinct chorion gene families, A and B, from the silkworm, B. mori, were able to express RNA in transformed D. melanogaster in a manner highly similar to what is seen in the silkworm moth.

The silkworm moth gene promoters directed expression in females during the late period of choriogenesis that takes place in the follicular epithelial cells of the ovaries. Therefore, despite the considerable evolutionary distance between the moths and flies, it is clear that some cis-acting features of the promoters are conserved. Furthermore, the evolutionary conservation of specific trans-acting protein is inferred. This study engendered much enthusiasm for using D. melanogaster as a host organism for assaying heterologous promoter function and first highlighted the possibility of deriving comparative information on the evolution of gene function. Further analyses by Mitsialis and colleagues (Mitsialis et al., 1987, 1989) defined a number of specific nucleotide sequences that were important for correct tissue-specific and stage-specific expression and for the abundance of the chorion gene transcription products. These studies highlighted the second reason for assaying promoters in a heterologous species, that is, the ability to ask questions about evolutionary conservation of both cis- and trans-acting features that regulate gene expression.

The initial work with the B. mori chorion genes was followed by another positive report of the proper function of a silkworm moth gene in D. melanogaster. In this case, the promoter of a gene, P25, encoding a silk protein, was linked to lacZ and used to transform the fly (Bello and Couble, 1990). The silk glands of larval moths are considered analogous to the larval salivary glands of the fly. Expression of the P25 promoter in transgenic flies resulted in detectable β-galactosidase activity in anterior larval salivary glands. Again, it was concluded that the cis-acting promoter sequences and trans-activating proteins were conserved over remarkable evolutionary distances in these two organisms. One difference in expression was noted, however. The P25 promoter normally expressed in the posterior cells of the moth silk gland, whereas it expressed in the anterior cells of the fly gland. These data indicated that although organ specificity was conserved, there was significant divergence in the location of putatively homologous cell types in these two species. In addition, they revealed an unexpected differentiation in the cell types of the fly larval salivary glands.

A series of studies with heterologous expression in D. melanogaster of promoters derived from midgut-specific genes provided further demonstration of the utility of cross-species transgenesis. The promoter of a gene encoding a carboxypeptidase from the black fly, Simulium vittatum, directs specific expression in the adult midgut of this species (Xiong and Jacobs-Lorena, 1995). Following transformation of D. melanogaster with a promoter fused with the β-glucoronidase reporter gene, expression was detected in larval and adult fly midguts. Although the tissue specificity was conserved, the temporal pattern of expression was different. Drosophila melanogaster appeared to express the gene constitutively, whereas the black fly showed significant induction during adult blood feeding. This difference in regulation perhaps is not surprising considering the significant physiological differences between a blood-feeding and non-blood-feeding insect.

Promoters derived from two midgut trypsin genes of the mosquito, *An. gambiae*, were linked to the *lacZ* gene and transformed into *D. melanogaster* (Skavdis *et al.*, 1996). Reporter gene expression was detected in the gut of both female and male adult flies. Interestingly, constructs made from the *AntrypI* gene, which is expressed relatively late in the mosquito digestive cycle (Müller *et al.*, 1995), were expressed throughout larval, pupal, and adult stages of the fly, although expression in first-instar larvae and pupae were low. In contrast, promoter fragments derived from another late trypsin gene, *Antryp2*, were weakly expressed in third-instar larvae, and pupae, and more strongly in adults. Although there are marked similarities of the primary structures of the putative promoter of the two genes, differences do clearly exist that are sufficient to affect temporal expression. Sequence analysis and protein-binding assays have been used to identify putative regulatory elements in the promoters of the trypsin genes (Shen and Jacobs-Lorena, 1998), but confirmation of the significance of the identified sequences requires the utilization of a homologous transformation system.

The analysis of hormone-inducible genes from sciarids indicates that features of ecdysone stimulation of gene transcription are conserved between *Sciara coprophila*, *Bradysia hygida*, and *D. melanogaster* (Bienz-Tadmor *et al.*, 1991; Monesi *et al.*, 1998). Furthermore, the *S. coprophila* gene was inducible following application of exogenous ecdysone to larval salivary glands dissected from transformed *D. melanogaster*. This implies that receptor mechanisms as well as cis- and trans-acting factors are conserved between higher diptera and nematocera. Interestingly, the temporal pattern of expression of the heterologous gene promoter did not follow the expected developmental profile of ecdysone abundance, indicating that species-specific fine-tuning of the expression pattern has evolved.

Aspects of the specificity of gene expression patterns in addition to timing also are not conserved in heterologous systems. For example, the *B. mori* chorion genes and sciarid hormone-inducible genes are amplified in copy number in the follicular cells and salivary glands, respectively. This amplification was not seen with the transgenes (Mitsialis and Kafatos, 1985; Bienz-Tadmor *et al.*, 1991; Shen and Jacobs-Lorena, 1998). Perhaps it is not too surprising that the DNA sequences responsible for gene amplification and developmentally regulated expression can be unlinked in these experiments. The gene amplification signals may act over large distances (Spradling and Orr-Weaver, 1987) and thus may not be included in the relatively small putative promoter fragments used in transgenic analyses. They may also be influenced by positional information that could be altered when a transgene inserts into a different region of the genome.

Sex-specific expression of genes also seems to be poorly conserved. For example, the *An. gambiae* trypsin genes are expressed in the adult female mosquitoes only, but following transgenesis they are expressed in both sexes of *D. melanogaster* (Müller *et al.*, 1995; Skavdis *et al.*, 1996). A similar uncoupling of sex specificity from tissue specificity was seen with *C. capitata*: the *MSSP-α2* gene is expressed only in male medflies, but is expressed in both sexes of *D. melanogaster* (Christophides *et al.*, 2000b). The same phenomenon has also been observed in genes expressed in the salivary glands of mosquitoes. The promoter of the *Apyrase* gene (*Agapy*) of *An. gambiae*, which is expressed in the adult salivary glands of female mosquitoes, was assayed in *D. melanogaster* (Lombardo *et al.*, 2000). While the *Agapy* promoter maintained its tissue and temporal expression patterns when introduced into *D. melanogaster*, it was expressed in the salivary glands of both females and males. The genetic basis of sex determination varies widely among the insects, even within members of the same order (Schutt and Nothiger, 2000); thus, the data we are seeing may reflect these fundamental differences in the importance of specific genes in the hierarchy of those that affect sex-specific gene expression.

Some of the most intriguing data come from genes that have been analyzed in both homologous and heterologous species. At the time this review was completed, three such studies exist. We have already described the results of the *MSSP* genes in *C. capitata*, where one gene, *MSSP-β2*, appears to be expressed the same in both species, but another and similar gene, *MSSP-α2*, has altered tissue and sex specificity. Differences in expression of the *Vg* gene promoter were observed when identical promoter–reporter gene constructs were introduced into *Ae. aegypti* and *D. melanogaster* (Kokoza *et al.*, 2000; A. S. Raikhel, personal communication). As discussed previously, the mosquito promoter was able to direct correct tissue-, stage-, and sex-specific expression when reintroduced back into the mosquito. However, when introduced into the fruit fly, this promoter was active in both the fat body and ovaries, as is true of the homologous fruit fly genes. This stands in contrast to what is observed in the mosquito, where this *Vg* gene is expressed only in the fat body. These results indicate that some features of the transcriptional control machinery that are present in the *Vg* gene are interpreted differently in the two species. The presence or absence of specific trans-activating factors in the different species is a possible explanation of these results, and this can result in an exogenous gene adopting a host-specific expression pattern. This observation is extended further by the analysis of the *Mal I* gene of *Ae. aegypti*. This gene is transcribed in the adult salivary glands of both male and female adult mosquitoes and therefore shows stage and tissue specificity (James *et al.*, 1989). Isolated putative promoter fragments can reproduce this expression pattern in transgenic mosquitoes (Coates *et al.*, 1999). However, when transformed into *D. melanogaster*, the *MalI* gene was expressed in the gut, salivary glands, and other tissues involved in the feeding stages of the fly (Pott, 1995), in a manner similar to the expression of the homologous genes of the fly (Snyder and Davidson, 1983).

In summary, there is considerable experimental power in being able to analyze promoter function in homologous and heterologous species. Aspects of sex-, stage-, and tissue-specific gene expression maybe dissected accurately, and the evolutionary origin of promoter elements inferred. On the other hand, the presence or absence of specific trans-acting factors may impart only a partial picture of promoter function if only the heterologous system is used. While the positive data (correct functioning in some manner of a promoter) can be evaluated following experiments in a heterologous species, the absence of a positive result does not mean that the promoter will not function appropriately in its homologous species. Researchers should be aware of this when basing conclusions on data from heterologous experiments alone.

VII. CONCLUDING REMARKS: GENERIC TECHNOLOGIES AND INSECT GENETICS

Readers of early reviews in the area of nondrosophilid transformation would have correctly concluded that there was a concerted effort to develop a generic gene transfer technology for these insects. These efforts were focused on the *P* element and, as discussed earlier, failed. However, the premise of developing generic technologies has been supported by many developments over the past several years. Researchers seeking to use genetic tools in nondrosophilid insects now have a set of techniques with which to ply their skills. Most conspicuous of these are the four transposable-element systems, from four different transposable-element families, that can be used genetically to transform these insects. Genetic transformants can be identified using genetic markers such as the fluorescent protein genes. Several *D. melanogaster* promoters appear to function correctly in those nondrosophilids in which they have been tested and provide the means to drive the expression of genes in these species. Other genetic tools, generic in nature, will most likely function in nondrosophilid species. Indeed, the FRT/FLP site-specific recombination system has been known to function correctly in *Ae. aegypti* for 10 years now (Morris *et al.*, 1991). There are no apparent reasons why more recent developments in homologous genetic recombination in *D. melanogaster* (Rong and Golic, 2000) and in RNAi technology in this species (Kennerdell and Carthew, 2000) cannot be applied successfully to nondrosophilid species. Simply put, the components of a successful genetic technology suitable for these important insect species are now available.

Challenges do, however, remain. As described previously, we need to understand how these transposable elements are regulated if they are to be used with maximum efficiency in the laboratory. We also need to understand the factors that influence their spread though genomes and populations if they are to be used as driving agents for genes in field populations of insects. We need more genetic markers in order to increase the versatility of genetic analyses we seek to deploy. Perhaps the biggest challenge is dealing, in a cost-effective way, with the biology of many of these insect species. These are not *Drosophila*, and the life cycles need to be handled in different ways. Single-pair matings are not always possible. With the possible exception of *Ae. aegypti* (in which eggs can be stored for up to 6 months), maintenance of the many genetic strains that are now being developed will require much labor and cost. Techniques are urgently needed for the cost-effective long-term storage of the many genetic strains that will be generated. In this regard, recent progress in the extension of cryopreservation strategies for embryos into nondrosophilid insects is encouraging (Leopold and Atkinson, 1999; Wang et al., 2000b). Finally, genetics offers many opportunities for solving major problems associated with medically and agriculturally important insects. The next decades promise a wealth of new discoveries that will advance basic knowledge as well as contribute to applied goals of insect control.

Acknowledgments

The authors thank their colleagues and collaborators for sharing unpublished data, and Lynn Olson for help in typing the manuscript. Original work presented here was supported by grants from the National Institutes of Health (AI45741), The University of California Mosquito Research Grants, the Californian Citrus Board, and the Californian Department of Food and Agriculture to P.W.A. the National Institutes of Health (GM48102) to David A. O'Brochta and P.W.A. and the National Institutes of Health (AI29746, AI44238, and AI44800), The John D. and Catherine T. MacArthur Foundation, and the Burroughs-Wellcome Fund to A.A.J.

References

Adams, M. D., et al. (2000). The genome sequence of Drosophila melanogaster. Science **287**, 2185–2195.

Allen, M. L., LeVesque, C. S., O'Brochta, D. A., and Atkinson, P. W. (2001). Stable germ-line transformation of Culex quinquefasciatus (Diptera: Culicidae). J. Med. Entomol. **38**, 701–710.

Atkinson, P. W., Warren, W. D., and O'Brochta, D. A. (1993). The hobo transposable element of Drosophila can be cross-mobilized in houseflies and excises like the Ac element of maize. Proc. Natl. Acad. Sci. USA **83**, 9693–9697.

Barthmaier, E., and Fyrberg, E. (1995). Monitoring development and pathology of Drosophila indirect flight muscles using green fluorescent protein. Dev. Biol. **169**, 770–774.

Bauser, C. A., Elick, T. A., and Fraser, M. J. (1999). Proteins from nuclear extracts of two lepidopteran cell lines recognize the ends of TTAA-specific transposons piggyBac and tagalong. Insect Mol. Biol. **8**, 223–230.

Beall, E. L., and Rio, D. C. (1996). Drosophila IRBP/Ku p70 corresponds to the mutagen-insensitive mus309 gene and is involved in P-element excision in vivo. Genes Dev. **10**, 921–923.

Beall, E. L., and Rio, D. C. (1997). Drosophila P-element is a novel site specific endonuclease. Genes Dev. **11**, 2137–2151.

Beall, E. L., and Rio, D. C. (1998). Transposase makes critical contacts with, and is stimulated by, single-stranded DNA at the P element termini in vitro. EMBO J. **17**, 2122–2136.

Bello, B., and Couble, P. (1990). Specific expression of a silk-encoding gene of Bombyx in the anterior salivary gland of Drosophila. Lett. Nature **346**, 480–482.

Berghammer, A. J., Klingler, M., and Wimmer, E. A. (1999). A universal marker for transgenic insects. Nature **402**, 370–371.

Bienz-Tadmor, B., Smith, H. S., and Gerbi, S. A. (1991). The promoter of DNA puff gene II/9-1 of Sciara coprophila is inducible by ecdysone in late prepupal salivary glands of Drosophila melanogaster. Cell Regul. **11**, 875–888.

Bigot, Y., Auge-Gouillou, C., and Periquet, G. (1996). Computer analyses reveal a *hobo*-like element in the nematode *Caenorhabditis elegans*, which presents a conserved transposase domain common with the Tc1-Mariner transposon family. *Gene* **174**, 265–271.

Blackman, R. K., Macy, M., Koehler, D., Grimaila, R., and Gelbart, W. M. (1989). dentification of a fully functional *hobo* transposable element and its use for germ line transformation of *Drosophila*. *EMBO J.* **8**, 211–217.

Brennan, M. D., Rowan, R. G., and Dickinson, W. J. (1984). Introduction of a functional *P* element into the germ-line of *Drosophila hawaiiensis*. *Cell* **38**, 147–151.

Callaerts, P., Halder, G., and Gehring, W. J. (1997). PAX-6 in development and evolution. *Annu. Rev. Neurosci.* **20**, 483–532.

Calvi, B. R., Hong, T. J., Findley, S. D., and Gelbart, W. M. (1991). Evidence for a common evolutionary origin of inverted terminal repeat transposons in *Drosophila* and plants: *Hobo*, *Activator* and *Tam3*. *Cell* **66**, 465–471.

Capy, P., Langin, T., Bigot, Y., Brunet, F., Daboussi, M. J., Periquet, G., David, J. R., and Hartl, D. L. (1994). Horizontal transmission versus ancient origin: *Mariner* in the witness box. *Genetica* **93**, 161–170.

Cary, L. C., Goebel, M., Corsaro, B. G., Wang, H. G., Rosen, E., and Fraser, M. J. (1989). Transposon mutagenesis of baculoviruses: analysis of *Trichoplusia ni* transposon IFP2 insertions within the FP-locus of nuclear polyhedrosis viruses. *Virology* **172**, 156–169.

Cathcart, L., Frafsur, E. S., Atkinson, P. W., and O'Brochta, D. A., *Hermes* distribution and structure in *Musca domestica*. Submitted.

Catteruccia, F., Nolan, T., Loukeris, T. G., Blass, C., Savakis, C., Kafatos, F. C., and Crisanti, A. (2000). Stable germline transformation of the malaria mosquito *Anopheles stephensi*. *Nature* **405**, 959–962.

Christophides, G. K., Livadaras, I., Savakis, C., and Komitopoulou, K. (2000a). Two medfly promoters that have originated by recent gene duplication drive distinct sex, tissue and developmental expression patterns. *Genetics* **156**, 173–182.

Christophides, G. K., Mintzas, A. C., and Komitopoulou, K: (2000b). Organization, evolution and expression of a multigene family encoding putative members of the odorant binding protein family in the medfly *Ceratitis capitata*. *Insect Mol. Biol.* **9**, 185–195.

Coates, C. J., Johnson, K. N., Perkins, H. D., Howells, A. J., O'Brochta, D. A., and Atkinson, P. W. (1996). The *hermit* transposable element of the Australian sheep blowfly, *Lucilia cuprina*, belongs to the *hAT* family of transposable elements. *Genetica* **97**, 23–31.

Coates, C. J., Turney, C. L., Frommer, M., O'Brochta, D. A., and Atkinson, P. W. (1997). Interplasmid transposition of the *mariner* transposable element in non-drosophilid insects. *Mol. Gen. Genet.* **253**, 728–733.

Coates, C. J., Jasinskiene, N., Miyashiro, L., and James, A. A. (1998). *Mariner* transposition and transformation of the yellow fever mosquito, *Aedes aegypti*. *Proc. Natl. Acad. Sci. USA* **95**, 3743–3747.

Coates, C. J., Jasinskiene, N., Pott, G. B., and James, A. A. (1999). Promoter-directed expression of recombinant fire-fly luciferase in the salivary glands of *Hermes*-transformed *Aedes aegypti*. *Gene* **226**, 317–325.

Coates, C. J, Jasinskiene, N., Morgan, D., Tosi, L. R. O., Beverley, S. M., and James, A. A. (2000). Purified *mariner* (*Mos1*) transposase catalyzes the integration of marked elements into the herm line of the yellow fever mosquito, *Aedes aegypti*. *Insect Biochem. Mol. Biol.* **30**, 1003–1008.

Corces, V., Pellicer, A., Axel, R., and Meselson, M. (1981). Integration, transcription, and control of a *Drosophila* heat shock gene in mouse cells. *Proc. Natl. Acad. Sci. USA* **78**, 7038–7042.

Cornel, A. J., Benedict, M. Q., Salazar-Rafferty, C., Howells, A. J., and Collins, F. H. (1998). Transient expression of the *Drosophila melanogaster cinnabar* gene rescues eye color in the white eye (WE) strain of *Aedes aegypti*. *Insect Biochem. Mol. Biol.* **27**, 993–997.

Craig, N. L. (1995). Unity in transposition reactions. *Science* **270**, 253–254.

Czerny, T., and Busslinger, M. (1995). DNA-binding and transactivation properties of Pax-6: Three amino acids in the paired domain are responsible for the different sequence recognition of Pax-6 and BSAP (Pax-5). *Mol. Cell. Biol.* **15**, 2858–2871.

Daniels, S. B., Strausbaugh, L. D., Ehrman, L., and Armstrong, R. (1984). Sequences homologous to *P* elements occur in *Drosophila paulistorum*. *Proc. Natl. Acad. Sci. USA* **81**, 6794–6797.

Daniels, S. B., Peterson, K. R., Strausbaugh, L. D., Kidwell, M. G., and Chovnick, A. (1990). Evidence for horizontal transmission of the *P* transposable element between *Drosophila* species. *Genetics* **124**, 339–355.

Elick, T. A., Bauser, C. A., Principe, N. M., and Fraser, M. J. (1995). PCR analysis of insertion site specificity, transcription, and structural uniformity of the Lepidopteran transposable element IFP2 in the TN-368 genome. *Genetica* **97**, 127–139.

Engels, W. R. (1989). *P* elements in *Drosophila*. *In* "Mobile DNA" (D. Berg and M. Howe, eds.). American Society for Microbiology, Washington, D.C.

Esposito, T., Gianfrancesco, F., Ciccodicola, A., Montanini, L., Mumm, S., D'Urso, M., and Forabosco, A. (1999). A novel pseudoautosomal human gene encodes a putative protein similar to *Ac*-like transposases. *Hum. Mol. Genet.* **8**, 61–67.

Fadool, J. M., Hartl, D. L., and Dowling, J. E. (1998). Transposition of the *mariner* element from *Drosophila mauritiana* in zebrafish. *Proc. Natl. Acad. Sci. USA* **95**, 5182–5186.

Fallon, A. M. (1991). DNA-mediated gene transfer: applications to mosquitoes. *Nature* **352**, 828–829.

Finnegan, D. J. (1985). Transposable elements in eukaryotes. *Int. Rev. Cytol.* **93**, 281–326.

Franz, G., and Savakis, C. (1991). *Minos*, a new transposable element from the *Drosophila hydei*, is a member of the *Tc1*-like family of transposons. *Nucleic Acids Res.* **19**, 6646.

Fraser, M. J., Smith, G. E., and Summers, M. D. (1983). Acquisition of host cell DNA sequences by baculoviruses: Relationship between host DNA insertions and FP mutants of *Autographa californica* and *Galleria mellonela* nuclear polyhedrosis viruses. *J. Virol.* **47**, 287–300.

Fraser, M. J., Ciszczon, T., Elick, T., and Bauser, C. (1996). Precise excision of TTAA-specific lepidopteran transposons *piggyBac* (IFP2) and *tagalong* (TFP3) from the baculovirus genome in cell lines from two species of Lepidoptera. *Insect Mol. Biol.* **5**, 141–151.

Garcia-Fernandez, J., Bayascas-Ramirez, J. R., Marfany, G., Munoz-Marmol, A. M., Casali, A., Baguna, J., and Salo, E. (1995). High copy number of highly similar mariner-like transposons in planarian (Platyhelminthe): Evidence for a trans-phyla horizontal transfer. *Mol. Biol. Evol.* **12**, 421–431.

Gomez-Gomez, E., Anaya, N., Roncero, M. I., and Hera, C. (1999). *Folyt1*, a new member of the *hAT* family, is active in the genome of the plant pathogen *Fusarium oxysporum*. *Fungal Genet. Biol.* **27**, 67–76.

Grappin, P., Audeon, C., Chupeau, M. C., and Grandbastien, M. A. (1996). Molecular and functional characterization of *Slide*, an *Ac*-like autonomous transposable element from tobacco. *Mol. Gen. Genet.* **252**, 386–397.

Grossman, G. L., Rafferty, C. S., Clayton, J. R., Stevens, T. K., Mukabayire, O., and Benedict, M. Q. (2002). Germ line transformation of the malaria vector, *Anopheles gambiae*, with the *piggyBac* transposable element. *Insect Mol. Biol.* **11** (in press).

Gueiros-Filho, F. J., and Beverley, S. M. (1997). Trans-kingdom transposition of the *Drosophila* element *mariner* within the protozoan *Leishmania*. *Science* **276**, 1716–1719.

Hagemann, S., Miller, W. J., and Pinsker, W. (1992). Identification of a complete *P*-element in the genome of *Drosophila bifasciata*. *Nucl. Acids Res.* **20**, 409–413.

Handler, A. M. (2000). An introduction to the history and methodology of insect gene transfer. *In* "Insect Transgenesis—Methods and Applications" (A. M. Handler and A. A. James, eds.), pp. 3–28. CRC Press, Boca Raton, FL.

Handler, A. M., and Gomez, S. P. (1996). The *hobo* transposable element excises and has related elements in tephritid species. *Genetics* **143**, 1339–1343.

Handler, A. M., and Gomez, S. P. (1997). A new *hobo, Ac, Tam3* transposable element, hopper, from *Bactrocera dorsalis* is distantly related to *hobo* and *Ac. Gene* **185,** 133–135.

Handler, A. M., and Harrell, III, R. A. (1999). Germline transformation of *Drosophila melanogaster* with the *piggyBac* transposon vector. *Insect Mol. Biol.* **8,** 449–457.

Handler, A. M., and Harrell, III, R. A. (2001). Transformation of the Caribbean fruit fly, *Anastrepha suspensa,* with a *piggyBac* vector marked with polyubiquitin-regulated GFP. *Insect Biochem. Mol. Biol.* **31,** 199–205.

Handler, A. M., and McCombs, S. M. (2000). The *piggyBac* transposon mediates germ-line transformation in the Oriental fruit fly and closely related elements exist in its genome. *Insect Mol. Biol.* **9,** 605–612.

Handler, A. M., McCombs, S. D., Fraser, M. J., and Saul, S. H. (1998). The lepidopteran transposon vector, *piggyBac,* mediates germ line transformation in the Mediterranean fruit fly. *Proc. Natl. Acad. Sci. USA* **95,** 7520–7525.

Haring, E., Hagemann, S., and Pinsker, W. (1995). Different evolutionary behavior of P element subfamilies: M-type and O-type elements in *Drosophila bifasciata* and *D. imaii. Gene* **163,** 197–202.

Hartl, D. L., Lohe, A. R., and Lozovskaya, E. R. (1997). Regulation of the transposable element mariner. *Genetica* **100,** 177–184.

Hediger, M., Niessen, M., Wimmer, E. A., Dubendorfer, A., and Bopp, D. (2001). Genetic transformation of the housefly *Musca domestica* with the lepidopteran derived transposon *piggyBac. Insect Mol. Biol.* **10,** 113–119.

Hehl, R., and Baker, B. (1990). Properties of the maize transposable element *Activator* in transgenic tobacco plants: A versatile inter-species genetic tool. *Plant Cell* **2,** 709–721.

Heinrich, J. C., Li, X., Henry, R. A., Haack, N., Stringfellow, L., Heath, A., and Scott, M. J. (2002). Germ line transformation of the Australian sheep blowfly *Lucilia cuprina. Insect Mol. Biol.* **11** (in press).

Hori, H., Suzuki, M., Inagaki, H., Oshima, T., and Koga, A. (1998). An active Ac-like transposable element in teleost fish. *J. Mar. Biotechnol.* **6,** 206–207.

Horn, C., and Wimmer, E. A. (2000). A versatile vector set for animal transgenesis. *Dev. Genes Evol.* **210,** 630–637.

Huttley, G. A., MacRae, A. F., and Clegg, M. T. (1995). Molecular evolution of the Ac/Ds transposable-element family in pearl millet and other grasses. *Genetics* **139,** 1411–1419.

James, A. A., Blackmer, K., and Racioppi, J. V. (1989). A salivary gland-specific, maltase-like gene of the vector mosquito, *Aedes aegypti. Gene* **75,** 73–83.

Jasinskiene, N., Coates, C. J., Benedict, M. Q., Cornel, A. J., Salazar-Rafferty, C., James, A. A., and Collins, F. H. (1998). Stable, transposon-mediated transformation of the yellow fever mosquito, *Aedes aegypti,* using the *Hermes* element from the house fly. *Proc. Natl. Acad. Sci. USA* **95,** 3743–3747.

Jasinskiene, N., Coates, C. J., and James, A. A. (2000). Structure of *Hermes* integrations in the germline of the yellow fever mosquito, *Aedes aegypti. Insect Mol. Biol.* **9,** 11–18.

Kaufman, P. D., and Rio, D. C. (1991). Germline transformation of *Drosophila melanogaster* by purified P element transposase. *Nucleic Acids Res.* **19,** 6336.

Kaufman, P. D., Doll, R. F., and Rio, D. C. (1989). *Drosophila* P element transposase recognizes internal P element sequences. *Cell* **59,** 359–371.

Kempken, F., and Kuck, U. (1996). Restless, an active Ac-like transposon from the fungus *Tolypocladium inflatum*: structure, expression, and alternative RNA splicing. *Mol. Cell. Biol.* **16,** 6563–6572.

Kennerdell, J. R., and Carthew, R. W. (2000). Heritable gene silencing in *Drosophila* using double-stranded RNA. *Nature Biotechnol.* **18,** 896–898.

Kidwell, M. G., Kidwell, J. F., and Sved, J. A. (1977). Hybrid dysgenesis in *Drosophila melanogaster:* A syndrome of aberrant traits including mutation, sterility, and male recombination. *Genetics* **86,** 813–833.

Klinakis, A. G., Loukeris, T. G., Pavlopoulos, A., and Savakis, C. (2000). Mobility assays confirm the broad host-range activity of the Minos transposable element and validate new transformation tools. *Insect Mol. Biol.* **9,** 269–275.

Kokoza, V., Ahmed, A., Cho, W. L., Jasinskiene, N., James, A. A., and Raikhel, A. (2000). Engineering blood meal-activated systemic immunity in the yellow fever mosquito, Aedes aegypti. *Proc. Natl. Acad. Sci. USA* **97,** 9144–9149.

Kokoza, V., Ahmed, A., Wimmer, E. A., and Raikhel, A. S. (2001). Efficient transformation of the yellow fever mosquito Aedes aegypti using the piggyBac transposable element vector, Bac [3xP3-EGFP afm]. *Insect Biochem. Mol. Biol.* **31,** 1137–1144.

Kusano, K., Johnson-Schlitz, D. M., and Engels, W. R. (2001). Sterility of Drosophila with mutations in the Bloom syndrome gene—Complementation by Ku70. *Science* **291,** 2600–2602.

Krebs, M. P., and Reznikoff, W. S. (1988). Use of a Tn5 derivative that creates lacZ translational fusions to obtain a transposition mutant. *Gene* **63,** 277–285.

Lampe, D. J., Churchill, M. E. A., and Robertson, H. M. (1996). A purified mariner transposase is sufficient to mediate transposition in vitro. *EMBO J.* **15,** 5470–5479.

Lampe, D. J., Akerley, B. J., Rubin, E. J., Mekalanos, J. J., and Robertson, H. M. (1999). Hyperactive transposase mutants of the Himar1 mariner transposon. *Proc. Natl. Acad. Sci. USA* **96,** 11428–11433.

Lampe, D. J., Walden, K. K. O., Sherwood, J. M., and Roberston, H. M. (2000). Genetic engineering of insects with mariner transposons. In "Insect Transgenesis—Methods and Applications" (A. M. Handler and A. A. James, eds.), pp. 237–248. CRC Press, Boca Raton, FL.

Lansman, R. A., Shade, R. O., Grigliatti, T. A., and Brock, H. W. (1987). Evolution of P transposable elements: Sequences of Drosophila nebulosa elements. *Proc. Natl. Acad. Sci. USA* **84,** 6491–6495.

Lee, S. H., Clark, J. B., and Kidwell, M. G. (1999). A P element-homologous sequence in the house fly, Musca domestica. *Insect Mol. Biol.* **8,** 491–500.

Lehane, M. J., Atkinson, P. W., and O'Brochta, D. A. (2000). Hermes-mediated genetic transformation of the stable fly, Stomoxys calcitrans. *Insect Mol. Biol.* **9,** 531–538.

Leopold, R. A., and Atkinson, P. W. (1999). Cryopreservation of sheep blow fly embryos, Lucilia cuprina (Diptera: Calliphoridae). *Cryo-Letters* **20,** 37–44.

Lidholm, D. A., Lohe, A. R., and Hartl, D. L. (1993). The transposable element mariner mediates germline transformation in Drosophila melanogaster. *Genetics* **134,** 859–868.

Lohe, A. R., and Hartl, D. L. (1996a). Autoregulation of mariner transposase activity by overproduction and dominant negative complementation. *Mol. Biol. Evol.* **13,** 659–555.

Lohe, A. R., and Hartl, D. L. (1996b). Germline transformation of Drosophila virilis with the transposable element mariner. *Genetics* **143,** 365–374.

Lohe, A. R., Moriyama, E. N., Lidhom, D. A., and Hartl, D. L. (1995). Horizontal transmission, vertical inactivation, and stochastic loss of mariner-like transposable elements. *Mol. Biol. Evol.* **12,** 62–72.

Lombardo, F., Di Cristina, M., Spanos, L., Louis, C, Coluzzi, M., and Arca, B. (2000). Promoter sequences of the putative Anopheles gambiae apyrase gene confer salivary gland expression in Drosophila melanogaster. *J. Biol. Chem.* **275,** 23861–23868.

Loukeris, T. G., Livadaras, I., Arca, B., Zabalou, S., and Savakis, C. (1995a). Gene transfer into the medfly, Ceratitis capitata, with a Drosophila hydei transposable element. *Science* **270,** 2002–2005.

Loukeris, T. G., Arca, B., Livadaras, I., Dialektaki, G., and Savakis, C. (1995b). Introduction of the transposable element Minos into the germ line of Drosophila melanogaster. *Proc. Natl. Acad. Sci. USA* **92,** 9485–9489.

Lozovskaya, E. R., Nurminsky, D. I., Hartl, D. L., and Sullivan, D. T. (1996). Germline transformation of Drosophila virilis mediated by the transposable element hobo. *Genetics* **142,** 173–177.

Martin, P., Martin, A., Osmani, A., and Sofer, W. (1986). A transient expression assay for tissue-specific gene expression of alcohol dehydrogenase in *Drosophila*. *Dev. Biol.* **117,** 574–580.

Maruyama, K., and Hartl, D. L. (1991). Evidence for interspecific transfer of the transposable element *mariner* between *Drosophila* and *Zaprionus*. *J. Mol. Evol.* **33,** 514–524.

Matsubara, T., Beeman, R. W., Shike, H., Besansky, N. J., Mukabayire, O., Higgs, S., James, A. A., and Burns, J. C. (1996). Pantropic retroviral vectors integrate and express in cells of the malaria mosquito, *Anopheles gambiae. Proc. Natl. Acad. Sci. USA* **93,** 6181–6185.

May, E. W., and Craig, N. L. (1996). Switching from cut-and-paste to replicative *Tn7* transposition. *Science* **272,** 401–404.

McClintock, B. (1950). The origin and behavior of mutable loci in maize. *Proc. Natl. Acad. Sci. USA* **36,** 344–355.

McGrane, V., Carlson, J. O., Miller, B. R., and Beaty, B. J. (1988). Microinjection of DNA into *Aedes triseriatus* ova and detection of integration. *Am J. Trop. Med Hyg.* **39,** 502–510.

Medhora, M., Maruyama, K., and Hartl, D. L. (1991). Molecular and functional analysis of the mariner mutator element *Mos1* in *Drosophila*. *Genetics* **128,** 311–318.

Michel, K., Stamenova, A., Pinkerton, A. C., Franz, G., Robinson, A. S., Gariou-Papalexiou, A., Zacharopoulou, A., O'Brochta, D. A., and Atkinson, P. W. (2001). *Hermes*-mediated germ-line transformation of the Mediterranean fruit fly, *Ceratitis capitata. Insect Mol. Biol.* **10,** 155–162.

Miller, L. H., Sakai, R. K., Romans, P., Gwadz, R. W., Kantoff, P., and Coon, H. G. (1987). Stable integration and expression of a bacterial gene in the mosquito, *Anopheles gambiae. Science* **237,** 779–781.

Miller, W. J., Hagemann, S., Reiter, E., and Pinsker, W. (1992). P-element homologous sequences are tandemly repeated in the genome of *Drosophila gauche. Proc. Natl. Acad. Sci. USA* **89,** 4018–4022.

Misra, S., and Rio, D. C. (1990). Cytotype control of *Drosophila* P element transposition: The 66 kd protein is a repressor of transposase activity. *Cell* **62,** 269–284.

Mitsialis, S. A., and Kafatos, F. C. (1985). Regulatory elements controlling chorion gene expression are conserved between flies and moths. *Nature* **317,** 453–456.

Mitsialis, S. A., Spoerel, N., Leviten, M., and Kafatos, F. C. (1987). A short 5'-flanking DNA region is sufficient for developmentally correct expression of moth chorion genes in *Drosophila. Dev. Biol.* **84,** 7987–7991.

Mitsialis, S. A., Veletza, S., and Kafatos, F. C. (1989). Transgenic regulation of moth chorion gene promoters in *Drosophila*: Tissue, temporal and quantitative control of four bidirectional promoters. *J. Mol. Evol.* **29,** 486–495.

Monesi, N., Jacobs-Lorena, M., and Paco-Larson, M. L. (1998). The DNA puff gene *BhC4-1* of *Bradysia hygida* is specifically transcribed in early prepupal salivary glands of *Drosophila melanogaster. Chromosoma* **107,** 559–569.

Moreira, L. A., Edwards, M., Adhami, F., Jasinskiene, N., James, A. A., and Jacobs-Lorena, M. (2000). Robust gut-specific gene expression in transgenic *Aedes aegypti* mosquitoes. *Proc. Natl. Acad. Sci. USA* **97,** 10895–10898.

Morris, A. C. (1997). Microinjection of mosquito embryos. *In* "The Molecular Biology of Insect Vectors of Disease" (J. M. Crampton, C. B. Beard, and C. Louis, eds.), pp. 423–429. Chapman and Hall, London.

Morris, A. C., Eggleston, P., and Crampton, J. M. (1989). Genetic transformation of the mosquito *Aedes aegypti* by micro-injection of DNA. *Med. Vet. Entomol.* **3,** 1–7.

Morris, A. C., Schaub, T. L., and James, A. A. (1991). FLP-mediated recombination in the vector mosquito, *Aedes aegypti. Nucleic Acids Res.* **19,** 5895–5900.

Morris, A. C., Pott, G. B., Chen, J., and James, A. A. (1995). Transient expression of a promoter-reporter construct in differentiated adult salivary glands and embryos of the mosquito *Aedes aegypti. Am. J. Trop. Med. Hyg.* **52,** 456–460.

Mullins, M. C., Rio, D. C., and Rubin, G. M. (1989). Cis-acting DNA sequence requirements for P-element transposition. *Genes Dev.* **3**, 729–738.

Müller, H. M., Catteruccia, F., Vizioli, J., della Torre, A., and Crisanti, A. (1995). Constitutive and blood meal-induced trypsin genes in *Anopheles gambiae. Exp. Parasitol.* **81**, 371–385.

O'Brochta, D. A., and Atkinson, P. W. Transformation Systems in Insects. *In* "Mobile Genetic Elements: Protocols and Genomic Applications." (W. J. Miller and P. Capy, eds.). Humana Press, Totowa, NJ (in press).

O'Brochta, D. A., Gomez, S. P., and Handler, A. M. (1991). P element excision in *Drosophila melanogaster* and related drosophilids. *Molec. Gen. Genet.* **225**, 3878–3894.

O'Brochta, D. A., Warren, W. D., Saville, K. J., and Atkinson, P. W. (1996). *Hermes,* a functional non-drosophilid gene vector from *Musca domestica. Genetics* **142**, 907–914.

Okuda, M., Ikeda, K., Namiki, F., Nishi, K., and Tsuge, T. (1998). *Tfo1:* An *Ac*-like transposon from the plant pathogenic fungus *Fusarium oxysporum. Mol. Gen. Genet.* **258**, 599–607.

Olson, K. E., Higgs, S., Gaines, P. J., Powers, A. M., Davis, B. S., Kamrud, K. I., Carlson, J. O., Blair, C. D., and Beaty, B. J. (1996). Genetically engineered resistance to dengue-2 virus transmission in mosquitoes. *Science* **272**, 884–886.

Paricio, N., Perez-Alonso, M., Martinez-Sebastian, M. J., and de Frutos, R. (1991). P sequences of *Drosophila subobscura* lack exon 3 and may encode a 66 kd repressor-like protein. *Nucleic Acids Res.* **19**, 6713–6718.

Paricio, N., Miller, W. J., Martinez-Sebastian, M. J., Hagemann, S., de Frutos, R., and Pinsker, W. (1996). Structure and organization of the P element related sequences in *Drosophila madeirensis. Genome* **39**, 823–829.

Peloquin, J. J., Thibault, S. T., and Miller, T. A. (2000). Genetic transformation of the pink bollworm *Pectinophora gossypiella* with the *piggyBac* element. *Insect Mol. Biol.* **9**, 323–333.

Perkins, H. D., and Howells, A. J. (1992). Genomic sequences with homology to the P element of *Drosophila melanogaster* occur in the blowfly *Lucilia cuprina. Proc. Natl. Acad. Sci. USA* **89**, 10753–10757.

Pinkerton, A. C., Whyard, S., Mende, H. M., Coates, C. J., O'Brochta, D. A., and Atkinson, P. W. (1999). The Queensland fruit fly, *Bactrocera tryoni,* contains mutiple members of the *hAT* family of transposable elements. *Insect Mol. Biol.* **8**, 423–434.

Pinkerton, A. C., Michel, K., O'Brochta, D. A., and Atkinson, P. W. (2000). Green fluorescent protein as a genetic marker in transgenic *Aedes aegypti. Insect Mol. Biol.* **9**, 1–10.

Plasterk, R. H., Izsvak, Z., and Ivics, Z. (1999). Resident aliens: The *Tc1/mariner* superfamily of transposable elements. *Trends Genet.* **15**, 326–332.

Pott, G. B. (1995). Characterization of salivary gland-specific control sequences of the mosquito, *Aedes aegypti.* Ph.D. thesis, University of California, Irvine.

Robertson, H. M. (1995). The *Tc1-mariner* superfamily of transposons in animals. *J. Insect. Physiol.* **41**, 99–105.

Robertson, H. M., and MacLeod, E. G. (1993). Five major subfamilies of *mariner* transposable elements in insects, including the Mediterranean fruit fly, and related arthropods. *Insect Mol. Biol.* **2**, 125–139.

Rong, Y. S., and Golic, K. G. (2000). Gene targeting by homologous recombination in *Drosophila. Science* **288**, 2013–2018.

Roth, C. R., Brey, P. T., Ke, Z., Collins, F. H., and Weissenbach, J. (2000). Direct submission of *Anopheles gambiae* STS clone 05H03 to GenBank. Accession number AL142367.

Rubin, E. J., Akerley, B. J., Novik, V. N., Lampe, D. J., Husson, R. N., and Mekalanos, J. J. (1999). *In vivo* transposition of *mariner*-based elements in enteric bacteria and mycobacteria. *Proc. Natl. Acad. Sci. USA* **96**, 1645–1650.

Rubin, G. M., and Spradling, A. C. (1982). Genetic transformation of *Drosophila* with transposable element vectors. *Science* **218**, 348–353.

Rubin, G. M., and Spradling, A. C. (1983). Vectors for P element-mediated gene transfer in *Drosophila*. *Nucleic Acids Res.* **11**, 6341–6351.

Sarkar, A., Coates, C. J., Whyard, S., Willhoeft, U., Atkinson, P. W., and O'Brochta, D. A. (1997a). The *Hermes* element from *Musca domestica* can transpose in four families of cylorrhaphan flies. *Genetica* **99**, 15–29.

Sarkar, A., Yardley, K., Atkinson, P. W., James, A. A., and O'Brochta, D. A. (1997b). Transposition of the *Hermes* element in embryos of the vector mosquito, *Aedes aegypti*. *Insect Biochem. Mol. Biol.* **27**, 359–363.

Scavarda, N. J., and Hartl, D. L. (1984). Interspecific DNA transformation in *Drosophila*. *Proc. Natl. Acad. Sci. USA* **81**, 7515–7519.

Schutt, C., and Nothiger, R. (2000). Structure, function and evolution of sex-determining systems in Dipteran insects. *Development* **127**, 667–677.

Shen, Z., and Jacobs-Lorena, M. (1998). Nuclear factor recognition sites in the gut-specific enhancer region of an *Anopheles gambiae* trypsin gene. *Insect. Biochem. Mol. Biol.* **28**, 1007–1012.

Sherman, A., Dawson, A., Mather, C., Gilhooley, H., Li, Y., Mitchell, R., Finnegan, D., and Sang, H. (1998). Transposition of the *Drosophila* element *mariner* into the chicken germ line. *Nature Biotechnol.* **16**, 1050–1053.

Shotkoski, F., Morris, A. C., James, A. A., and ffrench-Constant, R. H. (1996). Functional analysis of a mosquito gamma-aminobutyric acid receptor gene promoter. *Gene* **168**, 127–133.

Simonelig, M., and Anxolabehere, D. (1991). A P element of *Scaptomyza pallida* is active in *Drosophila melanogaster*. *Proc. Natl. Acad. Sci. USA* **88**, 6102–6106.

Skavdis, G., Siden-Kiamos, I., Müller, H. M., Crisanti, A., and Louis, C. (1996). Conserved function of *Anopheles gambiae* midgut-specific promoters in the fruitfly. *EMBO* **15**, 344–350.

Smartt, C. T., Kim, A. P., Grossman, G. L., and James, A. A. (1995). The Apyrase gene of the vector mosquito, *Aedes aegypti*, is expressed specifically in the adult female salivary glands. *Exp Parasitol.* **81**, 239–248.

Smit, A. F. A. (1999). Interspersed repeats and other mementos of transposable elements in mammalian genomes. *Curr. Opini. Genet. Dev.* **9**, 653–663.

Snyder, M., and Davidson, N. (1983). Two gene families clustered in a small region of the *Drosophila* genome. *J. Mol. Biol.* **166**, 101–118.

Spradling, A., and Orr-Weaver, T. (1987). Regulation of DNA replication during *Drosophila* development. *Annu. Rev. Genet.* **21**, 373–403.

Stark, K. R., and James, A. A. (1998). Isolation and characterization of the gene encoding a novel factor Xa-directed anticoagulant from the yellow fever mosquito, *Aedes aegypti*. *J. Biol. Chem.* **273**, 20802–20809.

Streck, R. D., MacGaffrey, J. E., and Beckendorf, S. K. (1986). The structure of *hobo* transposable elements and their insertion sites. *EMBO J.* **5**, 3615–3623.

Sundararajan, P., Atkinson, P. W., and O'Brochta, D. A. (1999). Transposable element interactions in insects: Crossmobilization of *hobo* and *Hermes*. *Insect Mol. Biol.* **8**, 359–368.

Tamura, T., Thibert, C., Royer, C., Kanda, T., Abraham, E., Kamba, M., Komoto, N., Thomas, J. L., Mauchamp, B., Chavancy, G., Shirk, P., Fraser, M., Prudhomme, J. C., Couble, P., Toshiki, T., Chantal, T., Corinne, R., Toshio, K., Eappen, A., Mari, K., Natuo, K., Jean-Luc, T., Bernard, M., Gerard, C., Paul, S., Malcolm, F., Jean-Claude, P., and Pierre, C. (2000). Germline transformation of the silkworm *Bombyx mori* L. using a *piggyBac* transposon-derived vector. *Nature Biotechnol.* **18**, 81–84.

Tsay, Y. F., Frank, M. J., Page, T., Dean, C., and Crawford, N. M. (1993). Identification of a mobile endogenous transposon in *Arabidopsis thaliana*. *Science* **260**, 342–344.

van Leunen, H. G. A. M., Colloms, S. D, and Plasterk, R. H. A. (1993). Mobilization of quiet endogenous Tc3 transposons of *Caenorhabditis elegans* by forced expression of Tc3 transposase. *EMBO J.* **12**, 2513–2520.

Wang, W., Swevers, L., and Iatrou, K. (2000a). *Mariner* (*Mos1*) transposase and genomic integration of foreign gene sequences in *Bombyx mori* cells. *Insect Mol. Biol.* **9,** 145–155.

Wang, W. B., Leopold, R. A., Nelson, D. R., and Freeman, T. P. (2000b). Cryopreservation of *Musca domestica* (Diptera: Muscidae) embryos. *Cryobiology* **41,** 153–166.

Warren, W. D., Atkinson, P. W., and O'Brochta, D. A. (1994). The *Hermes* transposable element from the housefly, *Musca domestica*, is a short inverted repeat-type element of the *hobo, Ac*, and *Tam3* (*hAT*) element family. *Genet. Res. (Camb.)* **64,** 87–97.

Warren, W. D., Atkinson, P. W., and O'Brochta, D. A. (1995). The Australian bushfly, *Musca vetustissima*, contains a sequence related to the transposons of the *hobo, Ac* and *Tam3* family. *Gene* **154,** 133–134.

Weaver, D. T. (1995). What to do to an end: DNA double-strand-break repair. *Trends Genet.* **11,** 388–392.

Witherspoon, T. J. (1999). Selective constraints on *P*-element evolution. *Mol. Biol. Evol.* **16,** 472–478.

Xiong, B., and Jacobs-Lorena, M. (1995). Gut-specific transcriptional regulatory elements of the carboxypeptidase gene are conserved between black flies and *Drosophila*. *Proc. Natl. Acad. Sci. USA* **92,** 9313–9317.

Yoshiyama, M., Honda, H., and Kimura, K. (2000). Successful transformation of the housefly, *Musca domestica* (Diptera: Muscidae) with the transposable element *mariner*. *Appl. Entomol. Zool.* **35,** 321–335.

Zhang, L., Sankar, U., Lampe, D. J., Robertson, H. M., and Graham, F. L. (1998). The *Himar1 mariner* transposase cloned in a recombinat adenovirus vector is functional in mammalian cells. *Nucleic Acids Res.* **26,** 3687–3693.

Zhao, Y.-G., and Eggleston, P. (1998). Stable transformation of an *Anopheles gambiae* cell line mediated by the *Hermes* mobile genetic element. *Insect Biochem. Mol. Biol.* **28,** 213–219.

Zhao, Y.-G., and Eggleston, P. (1999). Comparative analysis of promoters for transient gene expression in cultured mosquito cells. *Insect Mol. Biol.* **8,** 31–38.

3

Genes Mediating Sex-Specific Behaviors in *Drosophila*

Jean-Christophe Billeter, Stephen F. Goodwin, and Kevin M. C. O'Dell
IBLS Division of Molecular Genetics
University of Glasgow
Glasgow G11 6NU, United Kingdom

I. INTRODUCTION

Isolate a fertilized *Drosophila* egg prior to hatching, keep it in solitary confinement until it matures into an adult, and introduce it to a mature attractive fly of the opposite sex. Then, they mate. In fact they not only mate, they show the full

Copyright 2002, Elsevier Science (USA).
0065-2660/02 $35.00

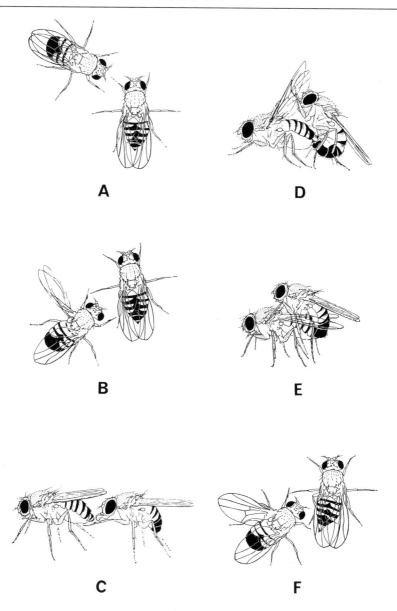

A

D

B

E

C

F

repertoire of courtship behaviors (Figure 3.1). If the ability to perform all aspects of sexual behavior can be achieved without parental guidance or advice from friends, then the capacity to perform courtship must be genetically determined.

However, before we get carried away with a genetic determinist agenda, we should appreciate that *Drosophila* sexual performance is affected by all sorts of environmental factors and is, in addition, modifiable by experience. For example, young males that have previously been courted by older males mate more quickly than males kept in isolation (McRobert and Tompkins, 1988). Males kept isolated since eclosion, so-called naive males, have been extensively employed to elucidate the role of learning in courtship behavior. For example, a naive male confined in a small chamber with an unreceptive female subsequently reduces the amount of courtship he directs during his next encounter, irrespective of the receptive state of the second female (Siegel and Hall, 1979; Gailey *et al.*, 1984). In addition, naive males also respond to sexually immature males with intense courtship that decreases markedly over a short period of time (Gailey *et al.*, 1982). Interestingly, males carrying mutations in previously described "learning and/or memory" genes fail to modify their behavior in light of previous sexual experience (Siegel and Hall, 1979). Indeed, the ability to learn and remember is thought to have evolved to increase reproductive fitness (Hall, 1986; Greenspan, 1995a). Nevertheless, it is clear that genes enable male and female *Drosophila* to perform and elicit sex-specific behavior. As *Drosophila* is a particularly tractable model genetic organism, it should be possible to identify genes that contribute to sex-specific behaviors and understand the molecular mechanisms by which they achieve this.

Despite numerous studies of *Drosophila melanogaster* courtship behavior (reviewed by Hall, 1994; Greenspan, 1995b; Yamamoto *et al.*, 1998; Goodwin, 1999; Greenspan and Ferveur, 2000), the precise mechanism by which genes actually control or determine sexual behavior is still relatively unclear (Baker *et al.*, 2001; Greenspan, 2001). The purpose of this review is to evaluate how genes mediate sex-specific behavior in *Drosophila*.

Figure 3.1. Courtship behavior in *Drosophila melanogaster*. Courtship involves a complex series of interactive behaviors. When presented with a female, a male (A) orientates toward the female and follows her when she moves, (B) shows unilateral wing vibration, (C) licks (via proboscis extension toward) the female's external genitalia, and (D) curls his abdomen toward her nether regions, which may result in (E) copulation (typically lasting 15–20 min for matings of this species). An unreceptive female will show (F) ovipositor extrusion. (Original figure from Burnet and Connolly, 1974, with permission.) ·

II. THE SINGLE-GENE APPROACH

Are there really fly courtship genes? Taken in its broadest sense all genes affect fly sexual behavior at some level or other (Grossfield, 1975; Ehrman, 1978). It is here that classical genetics, the single-gene approach of mutate and screen for a predetermined phenotype, suggest *Drosophila* as an ideal genetic organism for the study of behavior. Perturb the perception or generation of visual, auditory, or olfactory cues, and a male or female will usually mate at some reduced frequency (reviewed by Tompkins, 1984). Eliminate all three and you create a fly that is virtually behaviorally sterile. When paired with a receptive virgin wild-type female, a wild-type male has a courtship index (CI, percentage time spent courting; Siegel and Hall, 1979) of about 90%. However, males carrying both the *smellblind* and *glass* mutations, which eliminate his ability to smell and see, respectively, have a CI of just 5% (Tompkins, 1984). Similarly, when wing-clipped males were paired with virgin females in the dark (to preclude auditory and visual cues) the vast majority of pairs mated. However, when the males and females were additionally carrying the *olfactory-D* mutation (to preclude auditory, visual, and olfactory cues) only one pair in 17 mated (Gailey *et al.*, 1986). Thus, studies of sensory-impaired mutants reveal in some detail the significance of these nonsexual behaviors in courtship. However, this approach usually identifies genes specific to sensory perception or processing rather than genes whose primary function dictates fly sexual behavior (reviewed by Hall, 1994).

Our goal as neurogeneticists is to identify genes that influence sexual behavior without having a primary function in some aspect of nonsexual behavior. Such genes may be identified by isolating a mutant female that can hear, see, smell, and move normally, yet be hypo- or hyperreceptive to mating attempts. In the same spirit, we might discover a mutant male that may sing, hear, smell, see, and move normally, but be nondiscriminatory in terms of whom he courts. However, isolating sexual behavior mutants is not a trivial exercise. One successful strategy involves an initial screen for sterility, followed by a more detailed screen for aberrant courtship behavior. Mutants identified in this manner include the behaviorally sterile male mutant *celibate* (Hall *et al.*, 1980) and the hyporeceptive female mutant *icebox* (Kerr *et al.*, 1997). In this way we can distinguish between mutants whose sterility is associated with perturbations in innate courtship behavior as opposed to those that are sterile because of morphological defects in general reproductive organs (e.g., Castrillon *et al.*, 1993). A subset of behaviorally sterile males and females might be expected to have a courtship-specific defect that allows identification of a gene that plays a major role in determining sex-specific behavior.

In fact, few of the mutants isolated in screens for courtship defects specifically perturb courtship. Generally such mutations turn out to be pleiotropic, affecting a variety of sexual and nonsexual phenotypes (Hall, 1994; Greenspan, 2001). There are numerous examples of mutants isolated in a screen for one specific courtship defect that have subsequently been shown to have defects in a variety of behaviors. For example, the *cacophony* (*cac*) gene was originally identified in a screen for abnormal courtship song (von Schilcher, 1976) and was shown to encode a voltage-gated calcium channel $\alpha 1$-subunit (Smith *et al.*, 1996, 1998a). The gene itself is molecularly complex. The *cac* mRNA has multiple spice variants (Smith *et al.*, 1996; Peixoto *et al.*, 1997) and is subject to RNA editing (Smith *et al.*, 1998a,b). Independently isolated *cac* mutant alleles exhibit visual-system defects (*nightblind-A*: Heisenberg and Gotz, 1975; Smith *et al.*, 1998b), embryonic lethality (*l(1)L13*: Kulkarni and Hall, 1987) or temperature-sensitive paralysis (*cac*^{TS2}: Dellinger *et al.*, 2000).

Table 3.1. Mutant Genes That Perturb Courtship Behavior[a]

Mutant	Mutant Phenotype	Gene product
cacophony (*cac*)	Song defective	Calcium channel
celibate (*cel*)	Copulation defective	?
chaste (*cht*)	Female receptivity ↓	?
coitus interuptus (*coi*)	Copulation defective	?
courtless (*col*)	Courtship ↓	Ubiquitin-conjugating enzyme
croaker	Song defective	?
cuckold (*cuc*)	Courtship ↓	?
dissatisfaction (*dsf*)	Female receptivity ↓	"Tailless"-like nuclear receptor
dissonance (*diss*)	Song defective	RNA binding protein
don giovanni (*dg*)	Female courtship ↑	?
doublesex (*dsx*)	Song defective	Transcription factor
flamenco (*flam*)	Courtship absent	?
fickle (*fic*)	Copulation defective	Cytoplasmic tyrosine kinase
freeze (*fez*)	Blocked after orientation	?
fruitless (*fru*)	Orientation, song, and copulation defective	Transcription factor
he's not interested (*hni*)	Courtship ↓	?
icebox (*ibx*)	Female receptivity ↓	?
lingerer (*lig*)	Copulation defective	?
okina (*ok*)	Copulation defective	?
pale (*ple*)	Courtship ↓	Tyrosine hydroxylase
platonic (*plt*)	Copulation defective	?
quick-to-court (*qtc*)	Courtship ↑	Coiled-coil protein
spinster (*spin*)	Female receptivity ↓	Transmembrane protein
stuck (*sk*)	Copulation defective	?
tapered (*ta*)	Courtship ↓	?
yellow (*y*)	Courtship ↓	?

[a] The genetic factors listed here are a subset of those reported to affect aspects of courtship behavior in *Drosophila melanogaster*. Most are pleiotropic and hence not courtship-specific (↑) increase, (↓) decrease, (?) unknown. Most of the information currently available for *Drosophila* genes can be found at Flybase, the *D. melanogaster* Web site, at http://flybase.bio.indiana.edu/

To determine function, it is clearly critical to identify the molecular etiology underlying an aberrant behavioral phenotype. A common problem in dealing with mutant genes that perturb behavior is that phenotypes are rarely absolute, but are a matter of degree and may also show variations in penetrance. For example, males carrying the *stuck* mutation may mate normally or be unable to disengage from the female and die *in copula*, so *stuck*, ironically, has low penetrance. In the absence of a robust and absolute mutant phenotype it is often difficult to accurately map a mutation to a specific region of the *Drosophila* chromosome. Despite the completion of the *Drosophila* genome project (Kornberg and Krasnow, 2000), if we are incapable of mapping a mutation with some degree of accuracy, we cannot identify the relevant gene by positional cloning or subsequently ascribe molecular function.

It is probably fair to say that to date the single-gene approach has revealed surprisingly little about how genes mediate sex-specific behavior. However, we should not become disillusioned. Some of the genes listed in Table 3.1 will probably turn out to play critical roles in courtship, so we do not need to rename any of the genes "*he's not interesting.*"

From a more positive perspective, there are two classes of genes that largely fulfill our criteria of playing a critical role in the manifestation of sexual behavior without having a primary function in some aspect of nonsexual behavior. In this review, we will discuss the role of genes in the sex-determining cascade and those encoding male accessory gland peptides. In each case, specific aspects of sexual behavior may be explained in terms of the functions of molecules that can be ascribed to specific genes.

III. SEX DETERMINATION GENES DIRECT MORPHOLOGY, BIOCHEMISTRY, AND BEHAVIOR

Female and male *Drosophila melanogaster* differ in their appearance. Females are larger than males, have a narrower abdomen, and typically lack the darker abdominal pigmentation and the sex combs found in males. However, the matter of how flies differentiate as males or females extends far beyond these anatomical differences. Both sexes exhibit very different behaviors when it comes to reproduction.

The ability to perform sex-specific behavior must be dependent on sexual dimorphism in the fly brain. However, the brains of adult female and male *D. melanogaster* show very few anatomical differences. Indeed, sexually dimorphic anatomical features have only been found in two structures of the central nervous system (CNS). First, the mushroom bodies, which are central brain structures involved in olfactory learning and memory (Zars, 2000), have been shown to contain more neuronal fibers in females than in males (Technau, 1984). Second, the lobula in the optic lobes is slightly larger in males (Heisenberg *et al.*, 1995) and contains more neuropil than in females (Rein _et al._, 1999). The functional meaning of these differences in _D. melanogaster_ remains unknown. However, in larger flies (muscid and calliphorid) the lobula is a center for courtship tracking in males (Strausfeld, 1980). Nevertheless, no differences in anatomical organization between sexes have been detected in other prominent brain structures such as the antennal lobes (ALs) (Stocker _et al._, 1990; Laissue _et al._, 1999). The ALs are the primary olfactory association centers in insects and, in _D. melanogaster_, are innervated by neurons located in the antennae and in the maxillary palps (reviewed by Stocker, 1994). In contrast to _D. melanogaster_, females from some Hawaiian _Drosophila_ species have larger antennal lobes than males (Kondoh and Yamamoto, 1998) and sexually dimorphic ALs are found in several insects (e.g., Rospars, 1983; Hildebrand and Sheperd, 1997; Rospars and Hildebrand, 2000).

However, the absence of anatomical dimorphism in the majority of _D. melanogaster_ CNS structures does not preclude the absence of functional dimorphism. For example, genetic feminization of substructures in the ALs of an otherwise male fly leads to breakdown in mate recognition (Ferveur _et al._, 1995), showing the existence of a functional dimorphism of these structures between sexes. Hence, genes acting on the CNS to determine sexually dimorphic behaviors in _D. melanogaster_ probably act by influencing neuronal connectivity or changes in the brain chemistry and physiology rather than differences in gross anatomical structure.

The difference between sexes also involves biochemical cues. Mate recognition relies partially on sexually dimorphic and nondimorphic cuticular hydrocarbons that act as sex pheromones (reviewed by Ferveur, 1997; Greenspan and Ferveur, 2000). These pheromones are synthesised in subcuticular abdominal structures called oenocytes (Ferveur _et al._, 1997). The identified sex pheromones in _D. melanogaster_ are the predominant 7,11-heptacosadiene (7,11-HD) and 7,11-nonacosadiene (7,11-ND) in females and the predominant 7-tricosene (7-T) in males (reviewed by Jallon, 1984). The predominant female pheromones can induce male wing vibration (reviewed by Jallon, 1984) and the predominant male pheromones inhibit intermale courtship (Scott, 1986; Ferveur and Sureau, 1996). Until recently, these cuticular hydrocarbons were thought to be the exclusive pheromonal courtship-inducing cue.

However, genetic elimination of these known hydrocarbons suggests a revised picture for _Drosophila_ pheromonal action during courtship and in species isolation. Females whose cuticular hydrocarbons have been completely removed with an organic solvent become unattractive to males; however, females who only lack the female-specific pheromones (7,11-HD and 7,11-ND) remain attractive to males albeit at much lower levels (Savarit _et al._, 1999). Moreover, _D. melanogaster_ females are unattractive to males from other _Drosophila_ species, but genetic elimination of their female hydrocarbons leads to an increase in their attractiveness

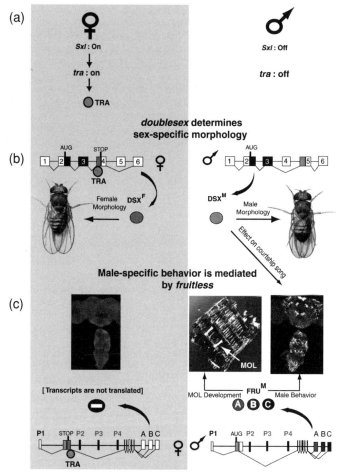

Figure 3.2. Genes determining sexual differentiation in *Drosophila melanogaster*. (a) In females the expression of the *Sex-lethal* (*Sxl*) gene controls the production of functional TRA protein by sex-specific splicing. In males *Sxl* is silent and *transformer* (*tra*) is default spliced to make a nonfunctional protein. (b) TRA controls the sex-specific splicing of *doublesex* (*dsx*) RNA; this gene determines most aspects of sex-specific development (see text). In males *dsx* pre-mRNA is default-spliced to generate the male-specific protein DSX^M that directs male development. In females TRA protein binds to the *dsx* pre-mRNA, contributing to the production of an alternative splice form that generates the female-specific DSX^F protein. Note that the two proteins have the exons 1, 2, 3, and 6 in common. However, females retain exon 4, whereas males retain exon 5. For details about the gene products shown, see text; for an extensive review, see Cline and Meyer (1996). (c) TRA also controls the sex-specific splicing of *fruitless* (*fru*), a gene that influences many features of male-specific behavior. In males, *fru* pre-mRNA from the P1 promoter is default-spliced to produce three alternative full-length proteins (FRU^M), each of which has exon A or

to males from other species. This study showed that the known *D. melanogaster* sex pheromones are not courtship activators as had been presumed for a long time, but rather modulators. Unknown pheromones ("ur-pheromones") should be present within the cuticle and act as male attractive substances both within and between *Drosophila* species. These pheromones await discovery.

The differentiation of male versus female external and internal morphology in *Drosophila* is dependent on genes of the sex-determination cascade (reviewed by Cline and Meyer, 1996). This cascade of genes also determines sex-specific behaviors by influencing neural development and function (reviewed by Goodwin, 1999; Greenspan and Ferveur, 2000).

The cascade of genes underlying morphological sex differences in *D. melanogaster* is well understood (Figure 3.2) and extensively reviewed (e.g., MacDougall *et al.*, 1995; Cline and Meyer, 1996). To make a long story short, if you are a female fly you express a functional Sex-lethal protein (Sxl), but if you are a male you do not express Sxl (Bell *et al.*, 1988). In females the Sxl protein regulates splicing of pre-mRNA transcribed from the *transformer* (*tra*) gene, so that an active Tra protein is expressed (Sosnowski *et al.*, 1989). Tra in combination with the non-sex-specifically expressed protein Transformer-2 (Amrein *et al.*, 1988) regulates the alternative splicing of at least two known targets; *doublesex* and *fruitless* (reviewed by Goodwin, 1999; Greenspan and Ferveur, 2000). If you are a male, neither Sxl nor Tra proteins are produced. Loss-of-function mutations in any of these genes convert chromosomally female (XX) into phenotypic males and expression of Sxl or Tra in males (XY) converts them to phenotypic females (see later discussion).

Several studies have shown a connection between the sex-determination hierarchy in *Drosophila* and the control of behavior. Males hemizygous for a deletion of the entire *Sxl* locus look, court, and copulate like wild-type males (Tompkins and McRobert, 1989). They synthesize the same sex pheromones as wild-type males and elicit very little courtship behavior from mature males, which is a normal attribute of a mature male's behavior. This implies that male behavior, morphology, and biochemistry occurs in the absence of *Sxl* activity. To test the effect of an active *Sxl* on male behavior, Tompkins and McRobert (1989) introduced a constitutively expressed temperature-sensitive allele of *Sxl* into chromosomally male flies. At permissive temperatures, chromosomal males are transformed into pseudo-females. When such males are raised at the restrictive temperature

---

B or C in a relatively 3'-region of the mature mRNA. These P1 transcripts are expressed in the male CNS to direct male-specific behavior and formation of the muscle of Lawrence (MOL). In females TRA protein binds to the *fru* pre-mRNA to produce transcripts that have a premature stop codon, so no functional FRU protein is produced. Note that transcripts (not shown) from promoters P2, P3, and P4 are found in both sexes and appear to be involved in *fru*'s vital function (Goodwin *et al.*, 2000; Anand *et al.*, 2001). (See color insert.)

of 25°C, *Sxl* expression becomes subnormal and these flies are phenotypically transformed into intersexual individuals. Such flies do not exhibit courtship when presented with a female but do elicit courtship from mature males. Analysis of the pheromonal profile of these intersexual males showed that they synthesize the predominant female aphrodisiac pheromone 7,11-HD (Tompkins and McRobert, 1989). A male fly expressing an aphrodisiac pheromone is also expected to be auto-elicited to court. The absence of such behavior in these intersexual males is further evidence that Sxl female function inhibits male behavior. This suggests that *Sxl* controls behavioral, morphological, and biochemical aspects of female sexual differentiation.

The connection between *tra* and sexual behavior has also been studied. A chromosomally female fly carrying a *tra* null mutation looks and behaves like a wild-type male (McRobert and Tompkins, 1985; Tompkins and McRobert, 1989) and is able to produce an almost normal male courtship song (Kyriacou and Hall, 1980; Kulkarni and Hall, 1987; Bernstein *et al.*, 1992). These phenotypic males do not produce characteristic female aphrodisiac pheromones and do not elicit courtship from wild-type males (Tompkins and McRobert, 1989). The Tra protein has also been shown to activate the production of female pheromones (Jallon *et al.*, 1986; Ferveur *et al.*, 1997). Therefore, mutations in genes upstream in the sex-determination hierarchy act on all aspects of male and female differentiation (reviewed by Hall, 1994).

A. Sex-specific morphology and biochemistry is primarily determined by *doublesex*

The *doublesex* gene produces functional proteins in both sexes, but these proteins are sexually dimorphic (Baker and Wolfner, 1988). In females *dsx* is Tra-spliced to make Dsx^F protein, whereas in males *dsx* is default spliced to produce Dsx^M (Hoshijima *et al.*, 1991; Ryner and Baker, 1991; see Figure 3.2). The two sex-specific Dsx proteins share a zinc-finger-related DNA binding domain, called the DM domain, at their N termini (Erdman and Burtis, 1993; Zhu *et al.*, 2000). Both proteins recognize and bind a consensus DNA sequence identified in the regulatory regions of the *yolk protein* genes (Burtis *et al.*, 1991; Coschigano and Wensink, 1993; An *et al.*, 1996; Erdman *et al.*, 1996). Also, within the common N-terminal domain is a subdomain that mediates homologous protein–protein interactions (An *et al.*, 1996; Erdman *et al.*, 1996; Cho and Wensink, 1997). The Dsx C-terminal domain is different between Dsx^M and Dsx^F and is believed to function as the sex-specific regulatory element through its own protein–protein interaction domain (Cho and Wensink, 1997).

The *dsx* gene is the last member in the hierarchy of regulatory genes required for several aspects of somatic sexual differentiation. Several genes and phenotypes have been directly or indirectly shown to be under the control of *dsx*. At the molecular level, only the *yolk protein* (*yp*) genes have been shown to be regulated by Dsx (Coschigano and Wensink, 1993; An and Wensink, 1995; Waterbury *et al.*, 1999). These genes control yolk-protein synthesis, which occurs in the ovaries and fat bodies of females, but not in males (Gelti-Douka *et al.*, 1974). Several features of the fly's sexual biochemistry have been shown to be perturbed in *dsx* mutants. The male-specific accessory gland peptides have been suggested to be under the control of *dsx* (Chapman and Wolfner, 1988). The predominant female pheromones (7,11-HD and 7,11-ND) are absent in *dsx* null mutants, and Dsx^F has been shown to induce their expression in males (McRobert and Tompkins, 1985; Jallon *et al.*, 1988; Waterbury *et al.*, 1999). In females Dsx^F acts both by activating genes required for female morphology and by repressing genes involved in male morphology. Dsx^F is sufficient to control the development of female internal and external attributes such as ovaries, genitalia, and abdominal pigmentation (Burtis and Baker, 1989; Szabad and Nothiger, 1992; Waterbury *et al.*, 1999). In males, Dsx^M activates genes required for male morphology and represses genes determining female morphology to generate a fly having male attributes including the sex-combs, the male-specific bristles on his forelegs (Jursnich and Burtis, 1993). The phenotypes of *dsx* mutant alleles demonstrate its involvement in the control of several aspects of the fly's internal and external sexual dimorphism, at both the morphological and biochemical levels.

Another gene critical for female-specific development is *intersex* (*ix*). In females, mutations in *ix* generate an intersexual fly that resembles a "*dsx*-like" mutant phenotype (McRobert and Tompkins, 1985). However, *ix* mutations have no effect in males (Baker and Ridge, 1980), showing that *ix* is necessary for normal female, but not male, development (Chase and Baker, 1995). This gene, which has yet to be cloned, is thought to be expressed in both sexes but only to affect females, probably through a direct interaction with the female-specific domain of Dsx^F (Waterbury *et al.*, 1999). The Ix protein is likely to act as a Dsx^F cofactor, enabling Dsx^F to positively activate at least *yp* expression and female pheromone synthesis (Waterbury *et al.*, 1999). Consistent with the absence of phenotypic change in *ix* males, such males exhibit the full courtship repertoire including copulation (Taylor *et al.*, 1994; Lee and Hall, 2000).

Although some might disagree ("Let's face it: looks are everything"—J. Nicholson, 1970), looks are not everything. Chromosomally female flies (XX) carrying a *dsx* allele (*dsx^{Dom}*) that constitutively expresses the male form of *dsx* look male but do not exhibit male behavior (Taylor *et al.*, 1994). Furthermore, a *dsx*-null XY male, though unable to copulate because "he" lacks external male genitalia, is still capable of performing some male courtship steps although at reduced levels (Villella and Hall, 1996; Waterbury *et al.*, 1999). Hence, Dsx^M is neither necessary nor sufficient for male courtship behavior. These findings strongly argue that after *tra* expression, sex-specific morphology and behavior are under quasi-independent genetic control. If sex-specific expression of Tra enables *dsx* to control morphological sex, then Tra presumably acts on one or more other genes to determine sex-specific behavior (Taylor *et al.*, 1994).

B. Male-specific behavior is primarily determined by *fruitless*

A candidate gene for the determination of sexual behavior is the male courtship mutant *fruitless* (*fru*: reviewed by Hall, 1994; Goodwin, 1999; Baker *et al.*, 2001). Mutant *fru* males court males and females indiscriminately, and populations of *fru* males form male–male courtship "chains" in which each male is courting and being courted simultaneously (Gailey and Hall, 1989; Villella *et al.*, 1997). Depending on the mutant allele, *fru* males show low levels of courtship, they sing abnormally or not at all, and they do not attempt copulation (Gailey and Hall, 1989; Ito *et al.*, 1996; Ryner *et al.*, 1996; Villella *et al.*, 1997; Goodwin *et al.*, 2000); or, if they succeed in mating (as males carrying certain *fru* mutations do), ejaculation problems ensue (Lee *et al.*, 2001). This last behavioral defect explains the sterility of *fru* males, which otherwise produce motile sperm (Gailey and Hall, 1989; Castrillon *et al.*, 1993). Morphologically, *fru* males appear to be normal, except for a male-specific pair of abdominal muscles called the muscle of Lawrence (MOL), which are incompletely formed or absent (Gailey *et al.*, 1991; Ito *et al.*, 1996; Villella *et al.*, 1997; Anand *et al.*, 2001). However, the absence of this structure in *fru* males is not the cause of their behavioral sterility because some MOL-less males are fertile (Gailey *et al.*, 1991; Villella *et al.*, 1997). This developmental defect has a neural etiology, as the formation of the MOL depends on its innervation by relevant genetically male motor neurons (Lawrence and Johnston, 1986; Usui-Aoki *et al.*, 2000). Furthermore, the MOL formation has been shown to be independent of *dsx* (Taylor, 1992). Mutations in *fru* appear to have no effect on female behavior (Villella *et al.*, 1997; Lee and Hall, 2000), although presumed null alleles are recessive lethal in both sexes (Anand *et al.*, 2001). This complex gene therefore has an essential function in both sexes, but a neurogenetic/behavioral role exclusive to males.

So does *fru* act on behalf of the nervous system and behavior in a manner that parallels the action of *dsx* mostly on other tissues insofar as their sexually dimorphic phenotypes are concerned? Molecular and functional analyses of *fru* reveal it to be a complex gene with multiple transcripts initiated from four promoters (P1-P4; Ryner *et al.*, 1996; see Figure 3.2). Transcripts from P1, the promoter located furthest away from the bulk of the *fru* coding sequence, are alternatively spliced by Tra (Ryner *et al.*, 1996; Heinrichs *et al.*, 1998). In females, Tra-mediated splicing introduces stop codons in the 5′ end of *fru* P1 transcripts and as a consequence, P1 Fru protein is undetectable (Lee *et al.*, 2000). In males, in the absence of Tra, *fru* pre-mRNAs are default-spliced to generate male-specific Fru proteins (Fru^M). Fru^M proteins (as well as the other Fru isoforms) have a BTB/POZ domain at the N terminus that has been implicated in protein dimerization, and three alternative zinc finger domains at their C terminus suggesting sequence-specific DNA binding (Ryner *et al.*, 1996; Ito *et al.*, 1996). Therefore Fru, like Dsx, is probably a transcription factor.

All the *fru* mutations that disrupt male behavior affect the sex-specific transcripts from the P1 promoter (Goodwin *et al.*, 2000; Lee and Hall, 2001). These transcripts are normally expressed and restricted to 1.7% of the neurons in the central nervous system from late third instar larva onward (Lee *et al.*, 2000; Lee and Hall, 2000; Usui-Aoki *et al.*, 2000). Transcripts from promoters P2, P3, and P4 are detected in both sexes in neural and nonneural tissues from embryonic stages onward and may encode *fru*'s vital function (Goodwin *et al.*, 2000; Lee *et al.*, 2000; Anand *et al.*, 2001). We speculate that in a male, genes regulated by one or all of the three different Fru^M isoforms cause the nervous system to develop in a male-specific fashion. It is possible that Fru^M isoforms either dimerize with each other through their BTB domain or form heterodimers with other BTB-containing proteins (e.g., the protein product of the *abrupt* gene, which is involved in axonal guidance; Hu *et al.*, 1995). This potential for combinatorial interaction could lead to the control of many downstream genes. A first candidate target of Fru^M is the gene (or genes) that determines male-specific serotonergic neurons as shown by colocalization of Fru^M and serotonin in the abdominal ganglion of adult male flies and the reduction in serotonin immunoreactivity in *fru* mutant males (Lee and Hall, 2001).

C. Interactions between *fruitless*, *doublesex*, and *dissatisfaction*

A hypothesis that Tra splices *dsx* to determine sexual morphology and splices *fru* to determine neurobehavioral sex is attractive. Both Dsx and Fru are putative transcription factors that may switch on the array of genes that would be necessary to achieve total "maleness" or total "femaleness" (Figure 3.2). However, this view is surely oversimplified. From a courtship behavioral perspective there is certainly functional crosstalk between these two genes. The male courtship song is disrupted, albeit in separate ways, by both *dsx* and *fru* (Villella and Hall, 1996; Villella *et al.*, 1997). Furthermore, Dsx^M has been shown to act on the formation of sexually dimorphic neuroblasts of the abdominal ganglion in males but not in females, showing that *dsx* also has a function in the central nervous system (Taylor and Truman, 1992).

Another instance of *dsx*-dependent behavioral phenotype is demonstrated by the *bric-a-brac* (*bab*) mutant. Females null for *bab* develop male abdominal pigmentation (Kopp *et al.*, 2000). This gene represses male-specific pigmentation in the last segments of the abdomen and is activated by Dsx^F (Kopp *et al.*, 2000). When presented with a choice between normal and *bab* mutant females, wild-type males discriminate strongly against dark pigmented *bab* mutant females. However, visually impaired *white* (*w*) mutant males court both females equivalently (Kopp *et al.*, 2000). Hence, visual cues such as female pigmentation also modulate male attraction and are controlled by *dsx*. The crosstalk between *fru* and *dsx* could also extend to a genetic interaction with *bab*. Indeed, although no genetic evidence show that *fru/bab* interact, it is attractive to speculate that *fru* and *bab* may have a physical interaction, since both encode proteins containing a BTB domain (Zollman *et al.*, 1994; Ito *et al.*, 1996; Ryner *et al.*, 1996), which is known to mediate protein–protein interactions (Albagli *et al.*, 1995).

The role of *dsx* in sexual behavior is episodically reassessed (Taylor *et al.*, 1994; Villella and Hall, 1996; Waterbury *et al.*, 1999). Interpreting the precise role of *dsx* is complicated by the fact that a gene determining sex-specific morphology and precious bodily fluids must have consequences for behavior. For example, the behavior of a fly that has a male brain and a female body might be expected to have an autoerotic phenotype, where its own pheromones act as an aphrodisiac. Furthermore, a male fly genetically induced to produce female pheromones elicits courtship from other males (Ferveur *et al.*, 1997).

There is some evidence that, as well as determining morphological differentiation, Dsx^F might also affect female behavior. Waterbury *et al.* (1999) made a transgene ubiquitously expressing Dsx^F and introduced this transgene into three classes of male (XY) flies. In wild-type (dsx^+/dsx^+) males, expression of Dsx^F generates flies that look and behave like males. However, these flies also produce female pheromones, so they are attractive to other males, whose advances they reject. In males missing one copy of *dsx* (dsx^+/dsx^-), expression of Dsx^F generates flies that show a lower frequency of courtship than control males (dsx^+/dsx^-) and have an intersexual morphology. They also produce female pheromones and again are attractive but unreceptive to other males. In *dsx*-null (dsx^-/dsx^-) mutant males, expression of Dsx^F generates flies that look like females, but they exhibit weak male behavior, directing infrequent courtship to females that does not go beyond the early stages. These flies are actively courted by other males and occasionally mate with them, but they generally attempt to reject such overtures both during courtship and copulation. Such behavior is reminiscent of mature XY: *dsx*⁻ mutants that elicit and reject courtship from wild-type males (Villella and Hall, 1996).

There are two plausible interpretations of these data. The first is that expression of Dsx^F in a male fly somehow interferes with the normal function of Fru^M. In this case, a high level of Dsx^F protein would be able to override the male behavioral state determined by Fru^M and direct some aspects of female behavior. This hypothesis requires a neural function for Dsx^F. A more attractive hypothesis is that, above certain relative concentrations, Dsx^F outcompetes Dsx^M (rather than Fru^M) and prevents the formation or differentiation of cells critical for males to sense stimuli from females (Villella and Hall, 1996; Waterbury *et al.*, 1999).

Males possess a variety of sexually dimorphic sensory structures throughout their peripheral nervous systems, including chemosensory sensillae in the pro- and mesothoracic legs and the maxillary palps (Nayak and Singh, 1983; Possidente and Murphey, 1989; Meunier *et al.*, 2000). The sex-specific development of these structures may be under the control of *dsx*. Under this scenario, constitutive expression of Dsx^F in *dsx*-null males (XY: *dsx*$^-$/*dsx*$^-$) generates a fly whose male central nervous system drives malelike courtship behavior (dependent on Fru^M). However, "he" has female-specific morphology, one feature of which is implied by "his" production of female pheromones from the relevant and relatively peripheral tissues (induced by Dsx^F) making him attractive to other males; and "he" critically lacks male-specific receptors that find female chemosensory cues attractive (repressed by Dsx^F), making "him" relatively unresponsive to females.

This hypothesis is supported by several other studies. Physical removal of the distal part of the first pair of legs in a male, used in tapping of the female abdomen during courtship, causes a breakdown in discrimination between female of his species and those of a closely related one, showing the involvement of this structure in mate recognition (Manning, 1959). The sensillae on the first pair of legs have been shown to be dimorphic not only in number but also in excitability to various chemicals (Meunier *et al.*, 2000). The neural projections from the leg sensilla to the thoracic ganglion are sexually dimorphic (Possidente and Murphey, 1989). According to a model in which *dsx* would control the differentiation of peripheral nervous structures, one could hypothesize that Fru^M-expressing neurons in the CNS are innervated by such structures. If these structures are female, they may fail to detect signals that would normally be sensed by male structures. It may then follow that the signal associated with the presence of a mature and receptive female would not be relayed to and processed by Fru^M-expressing neurons, such that the male behavioral program would not be triggered or could not be sustained. However, this hypothesis could only be supported if *dsx* mutations were to affect the differentiation of structures in the peripheral nervous system and if Fru^M neurons were somehow connected to these structures. At present this is pure conjecture. Nevertheless, Meunier *et al.* (2000) showed that in a male, the expression of the feminizing protein Tra^F in the leg nerves that are connected to sensillae involved in taste reception changes the sex specificity of these neurons. This is the first indication that a sex-determining gene can affect the sex-specific response of the peripheral nervous system to an external stimulus.

The question of precisely how female-specific behaviors are determined therefore remains unclear. Given that *fru* mutant males do not behave like females, in that they are not attractive or receptive to wild-type males (Villella *et al.*, 1997), *fru* determines male behavior without inhibiting female behavior. Therefore, genes that direct female-specific behavior presumably await discovery.

The story is further complicated by the discovery of *dissatisfaction* (*dsf*), another gene required for the manifestation of sexual behavior. Based on epistatic interactions, the *dissatisfaction* (*dsf*) gene appears to act downstream of *tra* in a *dsx*- and *fru*-independent manner. Mutant *dsf* males behave in a similar way to *fru* males, exhibiting indiscriminate courtship, quasi-chaining behaviors, and extended copulation duration (Finley *et al.*, 1997; Lee and Hall, 2000). Unlike *fru*, *dsf* also affects female behavior, causing her to be unreceptive to males and to lay immature eggs. The increased copulation duration of *dsf⁻* males and the egg-laying defect of *dsf⁻* females could be due to abnormal motoneuronal innervation of relevant muscles, so *dsf* would be required for sex-specific neural differentiation in both sexes (Finley *et al.*, 1997). The Dsf protein is related to the vertebrate Tailess/COUP class of nuclear receptor and contains both a DNA and ligand-binding domain. In the brain, *dsf* is expressed non-sex-specifically in a few subsets of neurons and does not appear to coexpress with *fru* (Finley *et al.*, 1998; Lee *et al.*, 2000). Although epistatic studies show *dsf* to be downstream of *tra* in the sex-determining hierarchy, the *dsf* mRNAs show no evidence of *tra*-mediated splicing. Consequently, Dsf is a non-sex-specific protein, which may interact, perhaps via its ligand-binding domain, with sex-specific partners to regulate its targets. In this fashion, *dsf* might be compared with *tra-2* and *ix*, whose sex specificity depends on interactions with their sex-specifically expressed partners—*tra* and *dsx*, respectively.

IV. THE BRAIN AS A SEXUAL ORGAN

Sex determination in *Drosophila* is cell-autonomous, allowing the creation of sex mosaics, flies that are part male and part female. An investigation of courtship behavior in sex mosaics implicates several regions of the CNS as being essential for male behavior (reviewed by Hall, 1994; Greenspan, 1995b). A molecular genetic approach can also generate sex mosaics. This enables targeted expression of the female-specific splice variant of *tra* to the male brain, to generate male flies that have specific domains of their brains genetically feminized (Ferveur *et al.*, 1995; O'Dell *et al.*, 1995; Ferveur and Greenspan, 1998). Interestingly, those brain domains identified in sex mosaics as having a critical role in controlling male-specific behavior coincide with regions of the brain that express *fru* sex-specific transcripts and Fru^M proteins (Ryner *et al.*, 1996; Ferveur and Greenspan, 1998; Goodwin, 1999; Goodwin *et al.*, 2000; Lee *et al.*, 2000; Usui-Aoki *et al.*, 2000). This suggests that Tra feminizes male behavior by splicing *fru* in the female mode in these subsets of cells (Ferveur *et al.*, 1995; O'Dell *et al.*, 1995; Ferveur and Greenspan, 1998; An *et al.*, 2000). However, interpreting the consequences of expressing *tra* in an otherwise male brain is not as simple as it may first appear. Critically, as discussed earlier, the behavioral function of *tra* may not solely be mediated by *fru*, but may in part be dependent on *dsx*.

Arthur *et al.* (1998) discovered a critical developmental period for the establishment of neurally controlled aspects of male sexual behavior. Time-controlled ubiquitous expression of Tra (leading to feminization) in males could only lead to loss of male behavior if it occurred no later than the early pupal stage (metamorphosis). This suggests that the sexual identity of the male CNS is determined during early metamorphosis and cannot be feminized after this period. Hence, the male behavioral program is "hard-wired" at a specific point during development. Consistent with this, Fru^M proteins that determine male-specific behavior are first expressed in the early pupal central nervous system (Lee *et al.*, 2000) and peak during a developmental time window overlapping with the putative phase of male behavioral wiring (Arthur *et al.*, 1998). We speculate that Fru^M proteins are actually responsible for the establishment of the male behavioral program during this period. Although these proteins are expressed at a time in development when significant neural reorganization is occurring (e.g., Truman, 1990; Ito and Hotta, 1992), they continue to be expressed throughout adulthood (Lee *et al.*, 2000), albeit at a lower level, suggesting a role in both initiation and maintenance of the male behavioral state. In this regard *fru*-induced absence or abnormalities of the MOL (Gailey *et al.*, 1991; Villella *et al.*, 1997) result from defects in muscle formation during development rather than a later degenerative effect. Conversely, in a *fru* male the absence of serotonin from specific abdominal ganglionic neurons that normally coexpress Fru^M (Lee and Hall, 2001, Lee *et al.*, 2001) does not seem to have a developmental etiology. In *fru* males, these neurons form and project their axons normally, but both subcellular compartments fail to produce serotonin. Therefore, *fru* functions encompass both developing and postdevelopmental stages of the life cycle.

Interpreting the courtship of transgenic flies, such as the *tra*-mediated sex mosaics described earlier, is further complicated by the unexpected ability of the transgene marker mini-*white* (which gives *D. melanogaster* its characteristic red eye color) to induce male–male courtship reminiscent of *fru* mutant males (Zhang and Odenwald, 1995; Hing and Carlson, 1996). However, expression of *tra* in the brain of a transgenic male in which the mini-*white* gene has been mutated still results in male–male courtship, implying that the change in sexual orientation is a direct consequence of *tra* expression (An *et al.*, 2000). In fact, males expressing mini-*white* can also mate with females and are fertile, so their behavior is quite distinct from that of most *fru* mutant male types. However, in the presence of a *fru* mutation that (by itself) causes near male/female "courtlessness," intermale courtships induced by mini-*white* were suppressed, suggesting *fru* and mini-*white* behavioral phenotype share a common neurogenetic pathway (Nilsson *et al.*, 2000).

V. THE ROLE OF SEROTONIN

Speculation that serotonin may play a role in *Drosophila* sexual orientation was made in light of the mini-*white* effect, in which ectopic and presumably ubiquitous expression of *w*-encoded protein causes males to court other males. At the molecular level, misexpression of *white*, which encodes a tryptophan/guanine transmembrane receptor, might also be expected to decrease levels of serotonin by lowering levels of its precursor tryptophan (Zhang and Odenwald, 1995). If a serotonin deficit induces male–male courtship, then *fru* and serotonin might be expected to colocalize to the same neurons. Antibodies raised against Fru^M and serotonin do not colocalize in the brain, which renders any link between *fru*, serotonin, and male sexual orientation more questionable. However, they do label *fru*-expressing neurons in the abdominal ganglion of adult males (Lee and Hall, 2001; Lee et al., 2001). These neurons send projections to the contractile muscle of the male's internal sex organs (Lee et al., 2001). In females and *fru* mutant males, these serotonergic neurons are putatively absent or do not express serotonin, respectively. Perhaps one downstream target of *fru* involves establishing the neurochemical differentiation of these abdominal serotonergic neurons.

Some combinations of *fru* alleles generate males that can mate, but the copulation duration of such males can be increased severalfold beyond the usual 15–20 minutes (Lee et al., 2001). The ability of such *fru* males to bend their abdomen and copulate in the absence of a muscle of Lawrence (MOL) shows this muscle to be unnecessary for copulation initiation. However, the extended copulation durations could be associated with a MOL defect, where a deficiency in MOL contractions (recall that the MOL is located in the dorsal part of the abdomen) would be expected to impinge on unbending and prevent the termination of copulation after a modest time span. However, whether a *fru* male that can mate completes mating within a reasonable period of time or after an extended one, he is frequently infertile because of a failure to transfer sperm, seminal fluid, or both (Lee et al., 2001).

The reproductive organs of these *fru* males appear morphologically normal and contain sperm, yet do not function normally. These organs are normally innervated by *fru* serotonergic neurons projecting from the abdominal ganglion (Lee et al., 2001). It is tempting to speculate that in a *fru* male, the serotonergic neurons that innervate the reproductive organs are nonfunctional, perhaps precluding contraction of muscles that surround the organs responsible for ejaculation (Lee et al., 2001).

VI. SEXUAL CONFLICT

At first sight, sexual encounters between male and female *Drosophila* are fleeting affairs, characterized by a few minutes of frenzied courtship, and a few or more minutes of copulation (depending on the species). For the male, at least, the consequences of mating on his subsequent behavior appear relatively trivial, as after a few moments' reflection he will actively pursue his next conquest (see Lee et al., 2001), a prisoner of his own innate promiscuity. For the female the effects of sex are rather more profound and life changing. Indeed, a mated female is an entirely different fly from a virgin, as within a few hours she undergoes a series of significant physiological and behavioral changes that may last for up to 10 days. She is relatively unreceptive and unattractive, her egg-laying rate increases dramatically, and her life expectancy is reduced (reviewed by Wolfner, 1997; Greenspan and Ferveur, 2000).

From an evolutionary perspective, it is initially difficult to see why it would be in the interest of recently mated females to undergo such a profound set of behavioral changes. However, the promiscuous male fly uses mating as an opportunity to protect his sperm investment. Mixed in with his sperm in the seminal fluid are the accessory-gland peptides, a cocktail of mind-altering "designer drugs" that act to maximize his reproductive success (Wolfner, 1997). His ex-partner is now unattractive and unreceptive, so is unlikely to mate with the competition. In addition, her egg-laying rate increases dramatically to make greatest use of his sperm. This is the chemical warfare department of sexual conflict.

There is also some evidence of sexual conflict between males that involves apparently aggressive head-to-head interactions (Lee and Hall, 2000). Such confrontations are exhibited at high frequency in males carrying either *fru* or *dsf* mutations, but are observed far less frequently in *dsx* or *ix* mutant males. The head-to-head interactions exhibited by *fru* males are reminiscent of putative aggressive behaviors exhibited by wild-type males and are functionally separable from courtship behavior itself (Lee and Hall, 2000).

A. The accessory-gland peptides

Seminal work from a number of laboratories identifies several accessory-gland peptides (Acp's: Table 3.2) that are synthesized by the male and passed to the female during copulation to change her postcopulatory behavior and physiology (reviewed by Wolfner, 1997). Expression of Acp genes is dependent on the male-specific splice-form of *dsx*, Dsx^M (Chapman and Wolfner, 1988). Some of the

Table 3.2. Accessory-Gland Proteins and Their Function[a]

Gene	Protein	Postcopulatory function in female
Acp26Aa	Prohormone-like peptide	Increased egg-laying and sperm displacement
Acp28Aa	Prohormone-like peptide	Stimulates oviposition
Acp29AB	Prohormone-like peptide	Sperm displacement
Acp36DE	Large glycoprotein	Sperm storage and sperm utilization
Acp53Eb	Simple peptide	Sperm displacement
Acp62F	Simple peptide	Seminal fluid toxicity?
Acp70A	Simple peptide	Increased egg-laying and decreased receptivity
Acp76A	Prohormone-like peptide	Seminal fluid coagulation?
Dup99B	Simple peptide	Increased egg-laying and decreased receptivity

[a]Details about Acp functions can be found in Wolfner (1997) and at Flybase, the *Drosophila* Web site, at http://flybase.bio.indiana.edu/

females that mate with semi-fertile *fru* males are left receptive to courtship, suggesting they do not receive seminal fluid from the mutant males (Lee *et al.*, 2001). The best-characterized accessory gland protein is Sex-Peptide (SP: Acp70A), a 36 amino-acid peptide, which increases egg-laying rate and decreases female sexual receptivity. Sex-Peptide's potency was demonstrated by Aigaki *et al.* (1991), who generated transgenic females carrying SP under heat-shock-promoter control. Within 2 hours of heat shock, these transgenic virgin females exhibit increased oviposition rate and decreased receptivity, which are definitive characteristics of the mated female state. Expression of a membrane-bound form of SP in a variety of tissues identifies a single target for SP that is restricted to the female brain (Nakayama *et al.*, 1997). However, a more recent study showed SP binding to a variety of tissues, including peripheral nerves, the suboesophageal ganglion, the cervical connective, discrete parts of the thoracic ganglion, and the genital tract (Ottiger *et al.*, 2000). Unequivocal identification of the SP target tissue(s) awaits the isolation of its cognate receptor.

The mechanism by which SP mediates these changes at the molecular level is now being revealed. Topical application of SP *in vitro* stimulates the biosynthesis of juvenile hormone from the corpora allata (CA) of female brains and activates vitellogenin uptake in maturing *Drosophila* oocytes (Soller *et al.*, 1999). Synthetic *Drosophila* SP also depresses sex pheromone production in the moth *Helicoverpa armigera* (Fan *et al.*, 1999). These effects can be achieved by the topical application of truncated forms of SP, the extreme N terminus being essential for CA stimulation, and the rest of the peptide depressing pheromone biosynthesis in a complex concentration-dependent manner (Fan *et al.*, 2000).

The postcopulatory changes in female attractiveness are multifactorial (Wolfner, 1997). There is a drop in the production of attractant pheromones, synthesis of antiaphrodisiacs, and acquisition of antiaphrodisiacs from her ex-partner. In addition, a mated female moves less than an otherwise equivalent virgin (Wolfner, 1997), and it is known that movement is a critical factor in stimulating males to initiate courtship (Tompkins *et al.*, 1982; Gailey *et al.*, 1986). Although the precise mechanism by which these postcopulatory behavioral changes occur has yet to be elucidated, it clearly requires specific Acp's to somehow modulate the function of appropriate regions of the female CNS. Within 5 minutes of the initiation of mating, specific Acp's have crossed the posterior vaginal intima and entered the hemolymph (Lung and Wolfner, 1999). There is no obvious common signal sequence or motif unique to those Acp's that enter the hemolymph (such as Acp26Aa and Acp62F) compared to those that do not (such as Acp36DE), and it is possible that transfer across the vaginal wall is size-selected (Lung and Wolfner, 1999). A more attractive hypothesis is that some Acp's are prevented from entering the hemolymph because they are tightly bound to sperm or to some part of the female's reproductive tract (Lung and Wolfner, 1999). One Acp that is retained in the female's reproductive tract (Acp36DE) is found in the sperm mass (Bertram *et al.*, 1996) tightly bound to sperm (Neubaum and Wolfner, 1999) and has been implicated in having a role in sperm storage (Chapman *et al.*, 2000). However, although the initiation of all these changes requires Acp's, maintenance of the "unattractive state" requires the presence of sperm (Tram and Wolfner, 1998).

The observation that mating reduces female lifespan implicates the Acp's as being in part responsible for increased female mortality (Chapman *et al.*, 1995; Harschman and Prout, 1994). Examination of age-specific mortality rates shows that mated females die faster than control virgin females at younger ages, but the rate becomes indistinguishable later in life (Civetta and Clark, 2000). Whether this is a consequence of accelerating egg-laying rate or a direct toxic effect of the male ejaculate is currently unclear. Interestingly, Acp62F exhibits significant sequence similarity to a spider neurotoxin (Wolfner *et al.*, 1997) and enters the hemolymph of females after mating (Lung and Wolfner, 1999). However, formal proof that Acp62F increases postmating female mortality has yet to be provided.

Further evidence that both Acp's and sperm are required to mediate all the postcopulatory changes in a female is provided by copulation complementation (Xue and Noll, 2000). Males carrying the *paired* mutation have no accessory glands; as a consequence they are sterile and do not decrease the postcopulatory sexual receptivity of their partner. However, if *paired*-mated receptive females are introduced to *tudor* males, which are spermless, the double-mated females produce viable offspring. Hence, independently providing a virgin female with sperm from a *paired* (ACP-null) mutant male and Acp's from a *tudor* (spermless) mutant male is sufficient to induce all the postcopulatory changes in her behavior described earlier (Xue and Noll, 2000).

From an evolutionary perspective, perhaps Acp's are not simply a selfish male tool designed to control female behavior. It is surely to her benefit, once having chosen to mate, to have her egg-laying rate determined by the presence of sperm, accompanied by oviposition-inducing Acp's. A mated female might wish to be left alone to lay eggs and otherwise conduct her affairs, such as eating. The alternative for the fertilized female of remating quickly is perhaps less appealing, as it not only wastes her time but makes her an easier target for predation. In fact, these male-induced changes in female physiology and behavior are neither permanent nor irreversible. After a period of about 11 days her egg-laying rate reduces, and she again becomes attractive and receptive and is likely to remate (Wolfner *et al.*, 1997).

B. Sperm competition

Evidence from wild-caught females suggests that most *Drosophila melanogaster* females mate with multiple males (Imhof *et al.*, 1998). The desire of the first-mating male to protect his sperm investment by switching off the female's receptivity is well founded, as in double-mated females around 80% of subsequent offspring are sired by the second male, a phenomenon called second-male sperm preference (Price *et al.*, 1999). Interestingly, if a sperm-carrying female subsequently mates with a spermless male, she produces fewer offspring than she would have done if she had not remated, implying a role for seminal fluid in second-male sperm preference (Gilchrist and Partridge, 1995). In a survey of sperm competitive ability, molecular variation in four Acp's (26A, 29B, 36DE, and 53E) was found to be significantly associated with the defensive sperm behavior, that is, the ability of sperm to resist being displaced by the ejaculate from a second male (Clark et al., 1995). However, we should appreciate that sperm competition is not an exclusively male-determined trait. There is some evidence that both remating frequency (Gromko and Newport, 1988) and degree of second male sperm preference (Clark and Begun, 1998) vary among different female genotypes.

A male with relatively low levels of Acp's transfers a slightly subnormal number of sperm to females, but only 10% of them are retained by her, suggesting a role for the Acp's in sperm storage (Tram and Wolfner, 1999). One Acp strongly implicated in sperm storage is Acp36DE, a protein that has been shown to bind directly to sperm. An *Acp36DE*-null male transfers sperm normally, but the female retains only 15% of his sperm, and only 10% of expected offspring are produced (Neubaum and Wolfner, 1999). In sperm competition experiments, *Acp36DE*-null males perform poorly, suggesting some kind of sperm-protecting role for Acp36DE protein (Chapman et al., 2000). Although efficient sperm storage has a critical role to play, the mechanism by which Acp36DE and other seminal fluid proteins determine sperm competition is unknown.

One other mechanism by which males appear to protect the health of their sperm is through the production of antibacterial proteins. Males produce at least three such proteins, one from their accessory glands and two from their ejaculatory duct. One of these appears to be andropin, a previously reported antibacterial peptide (Lung et al., 2001). During copulation the three peptides are passed to the female prior to the transfer of sperm, so they may have a sperm-protecting role within both male and female reproductive tracts.

There are a number of unanswered questions regarding sperm competition. In particular, it is unclear whether the sperm from two independent ejaculates are stored in distinct regions of the female's reproductive tract or whether they mix freely. Our understanding of the mechanisms underlying sperm competition would be greatly facilitated if we could visualize sperm within the reproductive system of a multiply mated female. Fortunately, there may be some light at the end of the tunnel. Transgenic males have been generated in which green-fluorescent protein (GFP) is driven by the sperm-specific promoter of the *don juan* gene to generate males that produce green fluorescing sperm (Price et al., 1999). The double-mating of females with normal and sperm-fluorescing males makes it clear that considerable mixing of the two ejaculates takes place. These data suggest that sperm competition involves both sperm displacement and sperm incapacitation, though the identity of the molecules that mediate these processes has yet to be revealed (Price et al., 1999).

VII. CONCLUSIONS

The completion of the *Drosophila* genome project in March 2000 inevitably gives us the sequence of every gene involved in courtship behavior. But how will we identify them? Will this speed up the discovery of courtship genes, or are we drowning in a sea of information without the necessary tools to identify them? Indeed, do courtship genes really exist, or are the gene interactions that enable a nervous system to develop and function in a sex-specific fashion so complex that few if any genes are specific to courtship?

In fact, most of the progress in identifying genes that mediate fly courtship has come from projects concerning previously described and well-understood biological phenomena, namely investigations of genes of the sex-determining cascade and genes encoding the accessory-gland peptides. Indeed, few novel mutations have so far been identified that might exhibit largely courtship-specific defects, or at least salient abnormalities of such behaviors among their other phenotypic problems. In this regard, it is worth bearing in mind that the arguably important story stemming from the discovery of *fru* involved the serendipitous recovery of the original mutant (Gill, 1963) in a screen for those being induced from a courtship-unrelated perspective (cf. Castrillon *et al.*, 1993). It is tempting to speculate that the next round of progress will come from identifying genes that act downstream of these previously described genes. Obvious questions are, how does *fru* wire-up a male-specific nervous system, and how do Acp's mediate the changes in female sexual receptivity? What happens to the function of her CNS after the males PBFs enter her nether regions? And how do intrinsic elements of her own neurochemistry act alongside, or interact with, the incoming, male-derived substances? Indeed, we hope that elucidation of these pathways is a finite distance beyond the horizon, so that all features of sex-specific neuronal connectivity within the CNS will be unraveled, and the meanings of sex-specific intraneuronal differentiations will become apparent.

Finally, what can *Drosophila* courtship genes tell us about human sexual behavior? Probably little or nothing, as few of these fly genes (*dsx* being a notable exception: Raymond *et al.*, 2000) have human counterparts. In any complex organism, brain formation and function must be controlled in part by genes. The question is, how does a given gene exert that control, so that its action directs some aspect of behavior? Even if a gene of this sort could be identified in a human, that does not mean it would solely "determine" behavior. By definition, those actions are also influenced by upbringing and environment, even in flies.

Acknowledgments

S.F.G is supported by grants from the Wellcome Trust, Royal Society, and Biotechnology and Biological Sciences Research Council; K.O.D is supported by a grant from the Biotechnology and Biological

Sciences Research Council; J.-C.B is supported by a University of Glasgow postgraduate scholarship and the Overseas Research Studentship award scheme. The authors thank Jeffrey C. Hall, Anthony Dornan, and Kim Kaiser for critical comments on the manuscript.

References

Aigaki, T., Fleischmann, I., Chen, P. S., and Kubli, E. (1991). Ectopic expression of sex peptide alters reproductive behavior of female *D. melanogaster*. *Neuron* **7**, 557–563.

Albagli, O., Dhordain, P., Dewindt, C., Lecocq, G., and Leprince, D. (1995). The BTB/POZ domain: A new domain protein-protein interaction motif common to DNA- and actin-binding protein. *Cell Growth Differ.* **6**, 1193–1198.

Amrein, H., Gorman, M., and Nothiger, R. (1988). The sex-determining gene *tra-2* of *Drosophila* encodes a putative RNA binding protein. *Cell* **55**, 1025–1035.

An, W., and Wensink, P. C. (1995). Three protein binding sites form an enhancer that regulates sex- and fat body-specific transcription of *Drosophila* yolk protein genes. *EMBO J.* **14**, 1221–1230.

An, W., Cho, S., Ishii, H., and Wensink, P. C. (1996). Sex-specific and non-sex-specific oligomerization domains in both of the *doublesex* transcription factors from *Drosophila melanogaster*. *Molec. Cell. Biol.* **16**, 3106–3111.

An, X., Armstrong, J. D., Kaiser, K., and O'Dell, K. M. C. (2000). The effects of ectopic *white* and *transformer* expression on *Drosophila* courtship behavior. *J. Neurogenet.* **14**, 227–243.

Anand, A., Villella, A., Ryner, L. C., Carlo, T., Goodwin, S. F., Song, H.-J., Gailey, D. A., Morales, A., Hall, J. C., Baker, B. S., and Taylor, B. J. (2001). Molecular genetic dissection of the sex-specific and vital functions of the *Drosophila melanogaster* sex determination gene *fruitless*. *Genetics* **158**, 1569–1595.

Arthur, B. I., Jallon, J.-M., Caflisch, B., Choffat, Y., and Nothiger, R. (1998). Sexual behaviour in *Drosophila* is irreversibly programmed during a critical period. *Curr. Biol.* **8**, 1187–1190.

Baker, B. S., and Ridge, K. A. (1980). Sex and the single cell. I. On the action of major loci affecting sex determination in *Drosophila melanogaster*. *Genetics* **94**, 383–423.

Baker, B. S., and Wolfner, M. F. (1988). A molecular analysis of *doublesex*, a bifunctional gene that controls both male and female sexual differentiation in *Drosophila melanogaster*. *Genes Devel.* **21**, 477–489.

Baker, B. S., Taylor, B. J., and Hall, J. C. (2001). Are complex behaviors specified by dedicated regulatory genes? Reasoning from *Drosophila*. *Cell* **105**, 13–24.

Bell, L. R., Maine, E. M., Schedl, P., and Cline, T. W. (1988). *Sex-lethal*, a *Drosophila* sex determination switch gene, exhibits sex-specific RNA splicing and sequence similarity to RNA binding protein. *Cell* **15**, 1037–1046.

Bernstein, A. S., Neumann, E. K., and Hall, J. C. (1992). Temporal analysis of tone pulses within the courtship songs of two sibling *Drosophila* species, their interspecific hybrid, and behavioral mutants of *Drosophila melanogaster* (*Diptera: Drosophilidae*). *J. Insect Behav.* **5**, 15–36.

Bertram, M. J., Neubaum, D. M., and Wolfner, M. F. (1996). Localization of the *Drosophila* male accessory gland protein Acp36DE in the mated female suggests a role in sperm storage. *Insect Biochem. Molec. Biol.* **26**, 971–980.

Burnet, B., and Connolly, K. J. (1974). Activity and sexual behavior in *Drosophila melanogaster*. In "The Genetics of Behavior" (van Abeelen, J. H. F., ed.), pp. 201–258. North Holland, Amsterdam.

Burtis, K. C., and Baker, B. S. (1989). *Drosophila doublesex* gene controls somatic sexual differentiation by producing alternatively spliced mRNAs encoding related sex-specific polypeptides. *Cell* **56**, 997–1010.

Burtis, K. C., Coschigano, K. T., Baker, B. S., and Wensink, P. C. (1991). The *doublesex* proteins of *Drosophila melanogaster* bind directly to a sex-specific yolk protein gene enhancer. *EMBO J.* **10**, 2577–2582.

Castrillon, D. H., Gonczy, P., Alexander, S., Rawson, R., Eberhart, C. G., Viswanathan, S., DiNaro, S., and Wasserman, S. A. (1993). Toward a molecular genetic analysis of spermatogenesis in *Drosophila melanogaster*: Characterization of male-sterile mutants generated by single *P*-element mutagenesis. *Genetics* **153**, 489–505.

Chapman, K. B., and Wolfner, M. F. (1988). Determination of male-specific gene expression in *Drosophila* accessory glands. *Devel. Biol.* **126**, 195–202.

Chapman, T., Liddle, L. F., Kalb, J. M., Wolner, M. F., and Partridge, L. (1995). Cost of mating in *Drosophila melanogaster* females is mediated by male accessory gland products. *Nature* **173**, 241–244.

Chapman, T., Neubaum, D. M., Wolfner, M. F., and Partridge, L. (2000). The role of male accessory gland protein Acp36DE in sperm competition in *Drosophila melanogaster*. *Proc. R. Soc. Lond. B* **267**, 1097–1105.

Chase, B. A., and Baker, B. S. (1995). A genetic analysis of *intersex*, a gene regulating sexual differentiation in *Drosophila melanogaster* females. *Genetics* **139**, 1649–1661.

Cho, S., and Wensink, P. C. (1997). DNA binding by the male and female *doublesex* proteins of *Drosophila melanogaster*. *J. Biol. Chem.* **272**, 3185–3189.

Civetta, A., and Clark, A. G. (2000). Correlated effects of sperm competition and postmating female mortality. *Proc. Natl. Acad. Sci. USA.* **97**, 13162–13165.

Clark, A. G., and Begun, D. J. (1998). Female genotypes affect sperm displacement in *Drosophila*. *Genetics* **149**, 1487–1493.

Clark, A. G., Aguade, M., Prout, T., Harschman, L. G., and Langley, C. H. (1995). Variation in sperm displacement and its association with accessory gland protein loci in *Drosophila melanogaster*. *Genetics* **139**, 189–201.

Cline, T. W., and Meyer, B. J. (1996). Vive la difference: Males vs females in flies vs worms. *Ann. Rev. Genet.* **30**, 637–702.

Coschigano, K. T., and Wensink, P. C. (1993). Sex-specific transcriptional regulation by the male and female *doublesex* proteins of *Drosophila*. *Genes Dev.* **7**, 42–54.

Dellinger, B., Felling, R., and Ordway, R. W. (2000). Genetic modifiers of the *Drosophila* NSF mutant, *comatose*, include a temperature-sensitive paralytic allele of the calcium channel $\alpha 1$-subunit gene, *cacophony*. *Genetics* **155**, 203–211.

Ehrman, L. (1978). Sexual behavior. *In* "The Genetics and Biology of *Drosophila*," Vol. 2b" (M. Ashburner and T. R. F. Wright, eds.), pp. 127–180. Academic Press, New York.

Erdman, S. E., and Burtis, K. C. (1993). The *Drosophila doublesex* proteins share a novel Zinc finger related DNA binding domain. *EMBO J.* **12**, 527–535.

Erdman, S. E., Chen, H. J., and Burtis, K. C. (1996). Functional and genetic characterization of the oligomerization and DNA binding properties of the *Drosophila doublesex* proteins. *Genetics* **144**, 1639–1652.

Fan, Y., Rafaeli, A., Gileadi, C., Kubli, E., and Applebaum, S. W. (1999). *Drosophila melanogaster* sex peptide stimulates juvenile hormone synthesis and depresses sex pheromone production in *Helicoverpa armigera*. *J. Insect Physiol.* **45**, 127–133.

Fan, Y., Rafaeli, A., Moshitzky, P., Kubli, E., Choffat, Y., and Applebaum, S. W. (2000). Common functional elements of *Drosophila melanogaster* seminal peptides involved in reproduction of *Drosophila melanogaster* and *Helicoverpa armigera* females. *Insect Biochem. Mol. Biol.* **30**, 805–812.

Ferveur, J.-F. (1997). The pheromonal role of cuticular hydrocarbons in *Drosophila melanogaster*. *Bioessays* **19**, 353–358.

Ferveur, J.-F., and Greenspan, R. (1998). Courtship behavior of brain mosaics in *Drosophila*. *J. Neurogenet.* **12**, 205–226.

Ferveur, J.-F., and Sureau, G. (1996). Simultaneous influence on male courtship of stimulatory and inhibitory pheromones produced live sex-mosaic *Drosophila melanogaster*. *Proc. R. Soc. Lond. B.* **263**, 967–973.

Ferveur, J.-F., Stoerkuhl, K. F., Stocker, R. F., and Greenspan, R. J. (1995). Genetic feminization of brain structures and changed sexual orientation in male *Drosophila*. *Science* **267**, 902–905.

Ferveur, J.-F., Savarit, F., O'Kane, C. J., Sureau, G., Greenspan, R. J., and Jallon, J.-M. (1997). Genetic feminization of pheromones and its behavioral consequences in *Drosophila* males. *Science* **276**, 1555–1558.

Finley, K. D., Taylor, B. J., Milstein, M., and McKeown, M. (1997). *dissatisfaction*, a gene involved in sex-specific behavior and neural development of *Drosophila melanogaster*. *Proc. Natl. Acad. Sci. USA.* **94**, 913–918.

Finley, K. D., Edeen, P. T., Foss, M., Gross, E., Ghbeish, N., Palmer, R. H., Taylor, B. J., and McKeown, M. (1998). *dissatisfaction* encodes a tailless-like nuclear receptor expressed in subset of CNS neurons controlling *Drosophila* sexual behavior. *Neuron* **21**, 1363–1374.

Gailey, D. A., and Hall, J. C. (1989). Behavior and cytogenetics of *fruitless* in *Drosophila melanogaster*: Different courtship defects caused by separate, closely linked lesions. *Genetics* **121**, 773–783.

Gailey, D. A., Jackson, F. R., and Siegel, R. W. (1982). Male courtship in *Drosophila*: The conditioned response to immature males and its genetic control. *Genetics* **102**, 771–782.

Gailey, D. A., Jackson, F. R., and Siegel, R. W. (1984). Conditioning mutations in *Drosophila* affect an experience-dependent courtship modification in courting males. *Genetics* **106**, 613–623.

Gailey, D. A., Lacaillade, R. C., and Hall, J. C. (1986). Chemosensory elements of courtship in normal and mutant, olfaction-deficient *Drosophila melanogaster*. *Behav. Genet.* **16**, 375–405.

Gailey, D. A., Taylor, B. J., and Hall, J. C. (1991). Elements of the *fruitless* locus regulate development of the muscle of Lawrence, a male-specific structure in the abdomen of *Drosophila melanogaster* adults. *Development* **113**, 879–890.

Gelti-Douka, H., Gingeras, T. R., and Kambysellis, M. P. (1974). Yolk proteins in *Drosophila*: Identification and site of synthesis. *J. Exp. Zool.* **187**, 167–72.

Gilchrist, A. S., and Partridge, L. (1995). Male identity and sperm displacement in *Drosophila melanogaster*. *J. Insect Physiol.* **41**, 1087–1092.

Gill, K. S. (1963). A mutation causing abnormal courtship and mating behavior in males of *Drosophila melanogaster* (abstract). *Am. Zool.* **3**, 507.

Goodwin, S. F. (1999). Molecular neurogenetics of sexual differentiation and behavior. *Curr. Opin. Neurobiol.* **9**, 759–765.

Goodwin, S. F., Taylor, B. J., Villella, A., Foss, M., Ryner, L. C., Baker, B. S., and Hall, J. C. (2000). Molecular defects in the expression of the *fruitless* gene of *Drosophila melanogaster* caused by aberrant splicing in P-element insertional mutants. *Genetics* **154**, 725–745.

Greenspan, R. J. (1995a). Flies, genes, learning, and memory. *Neuron* **15**, 747–750.

Greenspan, R. J. (1995b). Understanding the genetic construction of behavior. *Sci. Am.* **272(4)**, 74–79.

Greenspan, R. J. (2001). The flexible genome. *Nature Genet. Rev.* **2**, 383–387.

Greenspan, R. J., and Ferveur, J.-F. (2000). Courtship in *Drosophila*. *Ann. Rev. Genet.* **34**, 205–232.

Gromko, M. H., and Newport, M. E. A. (1988). Genetic basis of remating in *Drosophila melanogaster*. II. Response to selection based on the behavior of one sex. *Behav. Genet.* **18**, 621–632.

Grossfield, J. (1975). Behavioral mutants of *Drosophila*. In "Handbook of Genetics," Vol. 3 (R. C. King, ed.), pp. 679–702. Plenum, New York.

Hall, J. C. (1986). Learning and rhythms in courting, mutant *Drosophila*. *Trends Neurosci.* **9**, 414–418.

Hall, J. C. (1994). The mating of a fly. *Science* **264**, 1702–1714.

Hall, J. C., Siegel, R. W., Tompkins, L, and Kyriacou, C. P. (1980). Neurogenetics of courtship in *Drosophila*. *Stadler. Symp.* **12**, 43–82.

Harschman, L. G., and, Prout, T. (1994). Sperm displacement without sperm transfer in *Drosophila melanogaster*. *Evolution* **48**, 758–766.

Heinrichs, V., Ryner, L. C., and Baker, B. S. (1998). Regulation of sex-specific selection of *fruitless* 5' splice site by *transformer* and *transformer-2*. *Mol. Cell. Biol.* **18**, 450–458.

Heisenberg, M., and Gotz, K. G. (1975). The use of mutations for the partial degradation of vision in *Drosophila melanogaster*. *J. Comp. Physiol.* **98A**, 217–241.

Heisenberg, M., Heusipp, M., and Wanke, C. (1995). Structural plasticity in the *Drosophila* brain. *J. Neurosci.* **15**, 1951–1960.

Hildebrand, J. G., and Sheperd, G. M. (1997). Mechanisms of olfactory discrimination: Converging evidence for common principles across phyla. *Ann. Rev. Neurosci.* **20**, 595–631.

Hing, A. L., and Carlson, J. R. (1996). Male-male courtship behavior induced by ectopic expression of the *Drosophila white* gene: Role of sensory function and age. *J. Neurobiol.* **30**, 454–464.

Hoshijima, K., Inoue, K., Higushi, I., Sakamoto, H., and Shimura, Y. (1991). Control of *doublesex* alternative splicing by *transformer* and *transformer-2* in *Drosophila*. *Science* **252**, 833–836.

Hu, S., Fambrough, D., Atashi, J. R., Goodman, C. S., and Crews, S. T. (1995). The *Drosophila abrupt* gene encodes a BTB-zinc finger regulatory protein that controls the specificity of neuromuscular connections. *Genes Dev.* **9**, 2936–2948.

Imhof, M., Harr, B., Brem, G., and Schlotterer, C. (1998). Multiple mating in wild *Drosophila melanogaster* revisited by microsatellite analysis. *Mol. Ecol.* **7**, 915–917.

Ito, H., Fujitani, K., Usui, K., Shimizu-Nishikawa, K., Tanaka, S., and Yamamoto, D. (1996). Sexual orientation in *Drosophila* is altered by the *satori* mutation in the sex-determination gene *fruitless* that encodes a Zinc finger protein with a BTB domain. *Proc. Natl. Acad. Sci. USA.* **93**, 9687–9692.

Ito, K., and Hotta, Y. (1992). Proliferation of post-embryonic neuroblasts in the brain of *Drosophila melanogaster*. *Dev. Biol.* **149**, 134–148.

Jallon, J.-M. (1984). A few chemical words exchanged by *Drosophila* during courtship and mating. *Behav. Genet.* **14**, 441–478.

Jallon, J.-M., Antony, C., Chan-Yong, T. P., and Maniar, S. (1986). Genetic factors controlling the production of aphrodisiac substances in *Drosophila*. *In* "Advances in Invertebrate Reproduction," Vol. 4 (M. Porchet, J. C. Andrews, and A. Dhainaut, eds.), pp. 445–452. Elsevier, Amsterdam.

Jallon, J.-M., Lauge, G., Oessaud, L., and Antony, C. (1988). Female pheromones in *Drosophila melanogaster* are controlled by the *doublesex* locus. *Genet. Res.* **51**, 17–22.

Jursnich, V. A., and Burtis, K. C. (1993). A positive role in differentiation for the male *doublesex* protein of *Drosophila*. *Dev. Biol.* **155**, 235–249.

Kerr, C., Ringo, J., Dowse, H., and Johnson, E. (1997). *icebox,* a recessive X-linked mutation in *Drosophila* causing low sexual receptivity. *Behav. Genet.* **11**, 213–229.

Kondoh, Y., and Yamamoto, D. (1998). Sexual dimorphisms in the antennal lobes of two *Drosophila* species. *In* "Proceedings of the 26th Göttingen Neurobiology Conference" (G. Thieme, ed.), p. 388. Stuttgart. & New York.

Kopp, A., Duncan, I., and Carroll, S. B. (2000). Genetic control and evolution of sexually dimorphic characters in *Drosophila*. *Nature* **408**, 553–559.

Kornberg, T. B., and Krasnow, M. A. (2000). The *Drosophila* genome sequence: Implications for biology and medicine. *Science* **287**, 2218–2220.

Kulkarni, S. J., and Hall, J. C. (1987). Behavioral and cytogenetic analysis of the *cacophony* courtship song mutant and interacting genetic variants in *Drosophila melanogaster*. *Genetics* **115**, 461–475.

Kyriacou, C. P., and Hall, J. C. (1980). Circadian rhythm mutations in *Drosophila melanogaster* affect short-term fluctuations in the male's courtship song. *Proc. Natl. Acad. Sci. USA.* **77**, 6729–6733.

Laissue, P. P., Reiter, C., Hiesinger, P. R., Halter, S., Fischbach, K. F., and Stocker, R. F. (1999). Three-dimensional reconstruction of the antennal lobe in *Drosophila melanogaster*. *J. Comp. Neurol.* **405**, 543–552.

Lawrence, P. A., and Johnston, P. (1986). The muscle pattern of a segment of *Drosophila* may be determined by neurons and not by contributing myoblasts. *Cell* **45**, 505–513.

Lee, G., and Hall, J. C. (2000). A newly uncovered phenotype associated with the *fruitless* gene of *Drosophila melanogaster*: Aggression-like head interactions between mutant males. *Behav. Genet.* **30**, 263–275.

Lee, G., and Hall, J. C. (2001). Abnormalities of male-specific FRU protein and serotonin expression in the CNS of *fruitless* mutants in *Drosophila. J. Neurosci.* **21**, 513–126.

Lee, G., Foss, M., Goodwin, S. F., Carlo, T., Taylor, B. J., and Hall, J. C. (2000). Spatial, temporal, and sexually dimorphic expression patterns of the *fruitless* gene in the *Drosophila* central nervous system. *J. Neurobiol.* **43**, 404–426.

Lee, G., Villella, A., Taylor, B. J., and Hall, J. C. (2001). New reproductive anomalies in *fruitless*-mutant *Drosophila* males: Extreme lengthening of mating durations and infertility correlated with defective serotonergic innervation of reproductive organs. *J. Neurobiol.* **47**, 121–149.

Lung, O., and Wolfner, M. F. (1999). *Drosophila* seminal fluid proteins enter the circulatory system of the mated female fly by crossing the posterior vaginal wall. *Insect Biochem. Molec. Biol.* **29**, 1043–1052.

Lung, O., Kuo, L., and Wolfner, M. F. (2001). *Drosophila* males transfer antibacterial proteins from their accessory gland and ejaculatory duct to their mates. *J. Insect Physiol.* **47**, 617–622.

MacDougall, C., Harbison, D., and Bownes, M. (1995). The developmental consequences of alternate splicing in sex-determination and differentiation in *Drosophila. Devel. Biol.* **172**, 353–376.

Manning, A. (1959). The sexual isolation between *Drosophila melanogaster* and *Drosophila simulans. Anim. Behav.* **7**, 60–65.

McRobert, S. P., and Tompkins, L. (1985). The effect of *transformer, doublesex* and *intersex* mutations on the sexual behavior of *Drosophila melanogaster. Genetics* **111**, 89–96.

McRobert, S. P., and Tompkins, L. (1988). Two consequences of homosexual courtship performed by *Drosophila melanogaster* and *Drosophila affinis* males. *Evolution* **42**, 1093–1097.

Meunier, N., Ferveur, J.-F., and Marion-Poll, F. (2000). Sex-specific non-pheromonal taste receptors in *Drosophila. Curr. Biol.* **10**, 1583–1586.

Nakayama, S., Kaiser, K., and Aigaki, T. (1997). Ectopic expression of sex-peptide in a variety of tissues in *Drosophila* females using the P[GAL4] enhancer-trap system. *Mol. Gen. Genet.* **254**, 449–455.

Nayak, S. V., and Singh, R. N. (1983). Sensilla on the tarsal segments and mouthparts of adult *Drosophila melanogaster* Meigen (Diptera: Drosophilidae). *Int. J. Insect Morphol. Embryol.* **12**, 273–291.

Neubaum, D. M., and Wolfner, M. F. (1999). Mated *Drosophila melanogaster* females require a seminal fluid protein, Acp36DE, to store sperm efficiently. *Genetics* **153**, 845–857.

Nilsson, E. E., Asztalos, Z., Lukacsovich, T., Awano, W., Usui-Aoki, K., and Yamamoto, D. (2000). *fruitless* is in the regulatory pathway by which ectopic mini-*white* and *transformer* induce bisexual courtship in *Drosophila. J. Neurogenet.* **13**, 213–232.

O'Dell, K. M. C., Armstrong, J. D., Yang, M. Y., and Kaiser, K. (1995). Functional dissection of the *Drosophila* mushroom bodies by selective feminization of genetically defined subcompartments. *Neuron* **15**, 55–61.

Ottiger, M., Soller, M., Stocker, R. F., and Kubli, E. (2000). Binding sites of *Drosophila melanogaster* sex peptide pheromones. *J. Neurobiol.* **44**, 57–71.

Peixoto, A. A., Smith, L. A., and Hall, J. C. (1997). Genomic organization and evolution of alternative exons in a *Drosophila* calcium channel gene. *Genetics* **145**, 1003–1013.

Possidente, D. R., and Murphey, R. K. (1989). Genetic control of sexually dimorphic axon morphology in *Drosophila* sensory neurons. *Dev. Biol.* **132**, 448–457.

Price, C. S. C., Dyer, K. A., and Coyne, J. A. (1999). Sperm competition between *Drosophila* males involves both displacement and incapacitation. *Nature* **400**, 449–452.

Raymond, C. S., Murphy, M. W., O'Sullivan, M. G., Bardwell, V. J., and Zarkower, D. (2000). Dmrt1, a gene related to worm and fly sexual regulators, is required for mammalian testis differentiation. *Genes Dev.* **14**, 2587–2595.

Rein, K., Zockler, M., and Heisenberg, M. (1999). A quantitative three-dimensional model of the *Drosophila* optic lobes. *Curr. Biol.* **9**, 93–96.

Rospars, J. P. (1983). Invariance and sex-specific variations of the glomerular organization in the antennal lobes of a moth, *Mamestra brassicae*, and a butterfly, *Pieris brassicae*. *J. Comp. Neurol.* **220**, 80–96.

Rospars, J. P., and Hildebrand, J. G. (2000). Sexually dimorphic and isomorphic glomeruli in the antennal lobes of the sphinx moth *Manduca sexta*. *Chem. Senses* **25**, 119–129.

Ryner, L. C., and Baker, B. S. (1991). Regulation of *doublesex* pre-mRNA processing occurs by 3′-splice site activation. *Genes Devel.* **5**, 2071–2085.

Ryner, L. C., Goodwin, S. F., Castrillon, D. H., Anand, A., Villella, A., Baker, B. S., Hall, J. C., Taylor, B. J., and Wasserman, S. A. (1996). Control of male sexual behavior and sexual orientation in *Drosophila* by the *fruitless* gene. *Cell* **87**, 1079–1089.

Savarit, F., Sureau, G., Cobb, M., and Ferveur, J.-F. (1999). Genetic elimination of known pheromones reveals the fundamental chemical bases of mating and isolation in *Drosophila*. *Proc. Natl. Acad. Sci. USA.* **96**, 9015–9020.

Scott, D. (1986). Sexual mimicry regulates the attractiveness of mated *Drosophila melanogaster* females. *Proc. Natl. Acad. Sci. USA.* **83**, 8429–8433.

Siegel, R. W., and Hall, J. C. (1979). Conditioned responses in courtship behavior of normal and mutant *Drosophila*. *Proc. Natl. Acad. Sci. USA.* **76**, 3430–3434.

Smith, L. A., Wang, X., Peixoto, A. A., Neumann, E. K., Hall, L. M., and Hall, J. C. (1996). A *Drosophila* calcium channel alpha1 subunit gene maps to a genetic locus associated with behavioral and visual defects. *J. Neurosci.* **16**, 7868–7879.

Smith, L. A., Peixoto, A. A., and Hall, J. C. (1998a). RNA editing in the *Drosophila Dmca1A* calcium channel alpha 1 subunit transcript. *J. Neurogenet.* **12**, 227–240.

Smith, L. A., Peixoto, A. A., Kramer, E. M., Villella, A., and Hall, J. C. (1998b). Courtship and visual defects of *cacophony* mutants reveal functional complexity of a calcium channel α1 subunit in *Drosophila*. *Genetics* **149**, 1407–1426.

Soller, M., Bownes, M., and Kubli, E. (1999). Control of oocyte maturation in sexually mature *Drosophila* females. *Dev. Biol.* **208**, 337–351.

Sosnowski, B. A., Belote, J. M., and McKeown, M. (1989). Sex-specific alternative splicing of RNA from the *transformer* gene results from sequence-dependent splice site blockage. *Cell* **58**, 449–459.

Stocker, R. F. (1994). The organization of the chemosensory system in *Drosophila melanogaster*: A review. *Cell Tissue Res.* **275**, 3–26.

Stocker, R. F., Lienhard, M. C., Borst, A., and Fischbach, K. F. (1990). Neuronal architecture of the antennal lobe in *Drosophila melanogaster*. *Cell Tissue Res.* **262**, 9–34.

Strausfeld, N. J. (1980). Male and female visual neurones in Dipterous insects. *Nature* **283**, 381–383.

Szabad, J., and Nothiger, R. (1992). Gynandromorphs of *Drosophila* suggests one common primordium for the somatic cells of the female and male gonads in the region of abdominal segments 4 and 5. *Development* **115**, 527–533.

Taylor, B. J. (1992). Differentiation of a male-specific muscle in *Drosophila melanogaster* does not require the sex-determining genes *doublesex* or *intersex*. *Genetics* **132**, 179–191.

Taylor, B. J., and Truman, J. W. (1992). Commitment of abdominal neuroblasts in *Drosophila* to a male or female fate is dependent on genes of the sex-determining hierarchy. *Development* **114**, 625–642.

Taylor, B. J., Villella, A., Ryner, L. C., Baker, B. S., and Hall, J. C. (1994). Behavioral and neurobiological implications of sex-determining factors in *Drosophila*. *Dev. Genet.* **15**, 275–296.

Technau, G. M. (1984). Fiber number in the mushroom bodies of adult *Drosophila melanogaster* depends on age, sex and experience. *J. Neurogenet.* **1**, 113–126.

Tompkins, L. (1984). Genetic analysis of sex appeal in *Drosophila*. *Behav. Genet.* **14**, 411–440.

Tompkins, L., and McRobert, S. P. (1989). Regulation of behavioral and pheromonal aspects of sex determination in *Drosophila* by the *Sex-lethal* gene. *Genetics* **123**, 535–541.

Tompkins, L., Gross, A. C., Hall, J. C., Gailey, D. A., and Siegel, R. W. (1982). The role of female movement in the sexual behavior of *Drosophila melanogaster*. *Behav. Genet.* **12**, 295–307.

Tram, U., and Wolfner, M. F. (1998). Seminal fluid regulation of female sexual attractiveness in *Drosophila melanogaster*. *Proc. Natl. Acad. Sci. USA.* **95,** 4051–4054.

Tram, U., and Wolfner, M. F. (1999). Male seminal fluid proteins are essential for sperm storage in *Drosophila melanogaster*. *Genetics* **153,** 837–844.

Truman, J. W. (1990). Metamorphosis of the central nervous system of *Drosophila*. *J. Neurobiol.* **22,** 1072–1184.

Usui-Aoki, K., Ito, H., Ui-Tei, K., Takahashi, K., Lukacsovich, T., Awano, W., Nakata, H., Piao, Z. F., Nilsson, E. E., Tomida, J., and Yamamoto, D. (2000). Formation of the male-specific muscle in female *Drosophila* by ectopic *fruitless* expression. *Nat. Cell Biol.* **2,** 500–506.

Villella, A., and Hall, J. C. (1996). Courtship anomalies caused by *doublesex* mutations in *Drosophila melanogaster*. *Genetics* **143,** 331–344.

Villella, A., Gailey, D. A., Berwald, B., Ohshima, S., Barnes, P. T., and Hall, J. C. (1997). Extended reproductive role of the *fruitless* gene in *Drosophila melanogaster* revealed by behavioral analysis of new *fru* mutants. *Genetics* **147,** 1107–1130.

von Schilcher, F. (1976). The behavior of *cacophony*, a courtship song mutant in *Drosophila melanogaster*. *Behav. Biol.* **17,** 187–196.

Waterbury, J. A., Jackson, L. L., and Schedl, P. (1999). Analysis of the *doublesex* female protein in *Drosophila melanogaster:* Role in sexual differentiation and behavior and dependence on *intersex*. *Genetics* **152,** 1653–1667.

Wolfner, M. F. (1997). Tokens of love: Functions and regulation of *Drosophila* male accessory gland products. *Insect Biochem. Mol. Biol.* **27,** 179–192.

Wolfner, M. F., Harada, H. A., Bertram, M. J., Stelick, T. J., Kraus, K. W., Kalb, J. M., Lung, Y. O., Neubaum, D. M., Park, M., and Tram, U. (1997). New genes for male accessory gland proteins in *Drosophila melanogaster*. *Insect. Biochem. Mol. Biol.* **27,** 825–834.

Xue, L., and Noll, M. (2000). *Drosophila* female sexual behavior induced by sterile males showing copulation complementation. *Proc. Natl. Acad. Sci. USA.* **97,** 3272–3275.

Yamamoto, D., Fujitani, K., Usui, K., Ito, H., and Nakano, Y. (1998). From behavior to development: Genes for sexual behavior define the neuronal sexual switch in *Drosophila*. *Mech. Devel.* **73,** 135–146.

Zars, T. (2000). Behavioral functions of insect mushroom bodies. *Curr. Opin. Neurobiol.* **10,** 790–795.

Zhang, S. D., and Odenwald, W. F. (1995). Misexpression of the *white* (*w*) gene triggers male–male courtship in *Drosophila*. *Proc. Natl. Acad. Sci. USA.* **92,** 5525–5529.

Zhu, L., Wilken, J., Phillips, N. B., Narendra, U., Chan, G., Stratton, S. M., Kent, S. B., and Weiss, M. A. (2000). Sexual dimorphism in diverse metazoans is regulated by a novel class of intertwined zinc fingers. *Genes Dev.* **14,** 1750–1764.

Zollman, S., Godt, D., Prive, G. G., Couderc, J. L., and Laski, F. A. (1994). The BTB domain, found primarily in zinc finger proteins, defines an evolutionarily conserved family that includes several developmentally regulated genes in *Drosophila*. *Proc. Natl. Acad. Sci. USA.* **91,** 10717–10721.

4

Evolutionary Behavioral Genetics in *Drosophila*

Alexandre A. Peixoto
Department of Biochemistry and Molecular Biology
Fundação Oswaldo Cruz
CEP 21045-900
Rio de Janeiro, Brazil

ABSTRACT

Behavioral genes have a special evolutionary interest because they are potentially involved in speciation and in many forms of adaptation. Dozens of loci affecting different aspects of behavior have been already identified and cloned in *Drosophila*. Some of these genes determine variation in such ethologically complex phenotypes as the male "love song" that is produced during courtship and the locomotor "sleep–wake" activity cycles that are controlled by the circadian clock. Although the evolutionary analysis of most behavioral genes in *Drosophila* is relatively new, it has already given important insights into the forces shaping the molecular variation at these loci and their functional consequences. © 2002, Elsevier Science (USA).

Advances in Genetics, Vol. 47

I. INTRODUCTION

While the study of the function, regulation, and molecular evolution of developmental genes might prove to be fundamental to our understanding of macroevolution (e.g., Weatherbee et al., 1999), the evolutionary analysis of behavioral genes will turn out to be one of the most important approaches for studying the molecular basis of many aspects of speciation and adaptation.

Genes controlling sexual behavior, for example, are of particular interest in this respect, as they are likely to control species-specific differences in courtship that are involved in the reproductive isolation of closely related species (Coyne, 1992). Courtship in Drosophila melanogaster is characterized by a series of stereotyped behaviors that lead to copulation, and more than 30 genes have been identified through mutations that affect one or more of these elements (Hall, 1994a; Yamamoto et al., 1997; Greenspan and Ferveur, 2000). In addition, more than a dozen other genes have been identified by the analysis of their products, as in the case of the male accessory gland proteins (Wolfner, 1997). These "sexy" genes are potentially involved in the speciation process, arguably the major topic in evolution, and furthermore, it is also likely that they will be among the most common targets of sexual selection, a subject that has drawn special interest since Darwin (Andersson, 1994).

Behavioral genes also provide us with the opportunity to study the molecular basis of many types of adaptations. For example, the existence of an endogenous circadian ($\sim$24-hr) clock in most organisms is one of the most beautiful examples of the adaptation of life on the planet, in this case referring to the time span of the Earth's rotation (Pittendrigh, 1993). One wonders how conserved might be the genes that control this circadian clock, and whether their functions might be similar in different organisms. Other related questions concern differences among circadian systems and how they have evolved so that nocturnal and diurnal organisms retain the same underlying 24-hr rhythms while controlling their very different patterns of activity.

Genes that control aspects of behavior are likely to be pleiotropic (Hall, 1994b), so their evolution might be constrained or, in some cases driven by, the other phenotypes that they also determine. For the same reason, behavioral genes, such as those involved in courtship, might be under intense selective pressure (e.g., sexual selection related to female choice on male acoustic signals) that could cause changes in other phenotypes with associated long-term consequences.

The analysis of molecular variation within and between species will offer important clues about the forces shaping the evolution of behavioral genes. Here, I intend to discuss a few selected examples of work done in the field of evolutionary molecular biology of Drosophila behavioral genes, using examples from circadian rhythms and courtship. This review is not intended to be exhaustive, but rather to present and discuss what we have learned so far in these two exemplary evolutionarily genetic cases.

II. CLOCK GENES

The genetic analysis of biological rhythms was initiated by Konopka and Benzer (1971) when they isolated three mutations in D. *melanogaster* that speed up, slow down, or obliterate the circadian rhythms of locomotor activity and pupal–adult eclosion that are observed in wild-type flies. These mutations turned out to map to the same locus; it was accordingly anointed with the name of *period* (*per*) and was the first behavioral gene to be cloned and transformed in *Drosophila* (Bargiello and Young, 1984; Bargiello *et al.*, 1984; Reddy *et al.*, 1984; Zehring *et al.*, 1984). Since then, knowledge of the molecular basis for the *Drosophila* pacemaker has increased rapidly. The genetics and molecular analysis of *per* continues (Hamblen *et al.*, 1998), and several other clock genes have also been cloned and sequenced (reviewed by Hall, 1998; Young, 2000; Dunlap, 1999; Ripperger and Schibler, 2001).

The available evidence supports a model for the circadian pacemaker that involves the cyclical regulation of *per* and *timeless* (*tim*) gene transcription and translation via a negative feedback loop (Myers *et al.*, 1995; Hall, 1998; Dunlap, 1999; Young, 2000). Two further clock genes, *Clock* and *cycle*, encode proteins that have both a dimerization domain called PAS (also found in the PER protein) and a bHLH DNA-binding domain (Allada *et al.*, 1998; Darlington *et al.*, 1998; Rutila *et al.*, 1998). These transcription factors activate the cyclic expression of *per* and *tim* during the early part of the night (Hardin *et al.*, 1990; Sehgal *et al.*, 1995). As PER protein builds up in the cytoplasm, it is phosphorylated by the DOUBLETIME (DBT) casein kinase (Kloss *et al.*, 1998) and earmarked for degradation (Price *et al.*, 1998). As TIM builds up, it blocks the action of DBT, allowing PER levels to accumulate (Price *et al.*, 1998). Late at night the concentration of PER and TIM in the cytoplasm allows them to dimerize via the PER PAS domain and enter the nucleus of pacemaker cells (Curtin *et al.*, 1995; Saez and Young, 1996). This nuclear translocation is also affected by the phosphorylation of TIM by SGG/GSK-3 (Martinek *et al.*, 2001). The PER/TIM complex (probably via the PER PAS domain) interacts with the CLOCK/CYCLE dimer, repressing the transcription of *per* and *tim*. In the early morning, the blue light sensor CRYPTOCHROME (CRY) (Emery *et al.*, 1998; Stanewsky *et al.*, 1998) interacts with TIM and derepresses the CLOCK-CYCLE dimer, which is now free to promote transcription of *per* and *tim* (Ceriani *et al.*, 1999). Thus the negative feedback loop is complete.

The great majority of evolutionary studies of clock genes involves *per*, whereas such analysis of other genes is still in the early stages.

A. *period*

Evolutionary analyses of the *per* gene in *Drosophila* and closely related Diptera (Colot *et al.*, 1988, Peixoto *et al.*, 1992, 1993; Nielsen *et al.*, 1994; Piccin *et al.*,

2000; Warman *et al.*, 2000) have revealed that *per* has one of the fastest-evolving insect coding sequences. This rapid evolutionary rate appears to have important functional consequences. In a study of 26 lepidopteran species (Regier *et al.*, 1998), a rapid evolution of the PAS/CLD (PAS/cytoplasmatic localization domain) was revealed despite the fact that this is one of the most conserved regions within the *Drosophila*. The cloning and expression studies of a silk moth *period* gene homolog also revealed important differences in gene regulation compared with *Drosophila* that might be relevant to this sequence evolution, particularly as PER protein did not appear to go nuclear in brain cells (Reppert *et al.*, 1994; Sauman and Reppert, 1996; Gotter *et al.*, 1999).

Nearer home, the analysis of the *period* gene in *Musca domestica* revealed an unexpectedly high amino acid sequence similarity to *D. melanogaster* in a region that includes the PAS/CLD domain (Piccin *et al.*, 2000). This seemed to account for the better rescue of transformed per^{01} mutants observed with the *M. domestica* gene (Piccin *et al.*, 2000) compared with that obtained with *D. pseudoobscura*'s (Petersen *et al.*, 1988). The long-awaited cloning of mammalian homologs revealed the existence of three different *per* genes with distinct temporal and expression patterns (e.g., Zylka *et al.*, 1998; Zheng *et al.*, 2001).

One of the most striking features of the *per* gene in *D. melanogaster* is a repetitive region encoding several threonine–glycine (Thr-Gly) pairs (Jackson *et al.*, 1986; Citri *et al.*, 1987). The first indication that this region of *per* would be evolutionarily interesting came from the observation that laboratory strains of *D. melanogaster* show variation in the number of Thr-Gly pairs (Yu *et al.*, 1987). This prompted Costa *et al.* (1991) to examine natural populations of this species to assess whether this length variation among laboratory stocks reflected the existence of a polymorphism in the wild. The results indicated that European populations of *D. melanogaster* are highly polymorphic in the number of Thr-Gly pairs, ranging from 14 to 24 repeats (Costa *et al.*, 1991; Sawyer *et al.*, 1997). Moreover, analysis of the DNA sequences encoding the repeats revealed the likely origin of the different Thr-Gly alleles through a series of insertion and deletion events caused by processes such as replication slippage and unequal crossing-over (Costa *et al.*, 1991) (Figure 4.1, see color insert). In fact, when a larger number of populations were studied and more length alleles were sequenced, other variants were found, and a more complicated picture emerged (Rosato *et al.*, 1996). An interesting observation is that the great majority of the length variants differ by three Thr-Gly pairs (14, 17, 20, and 23). It turns out that the (Thr-Gly)$_3$ motif corresponds to a conformational monomer generating a β-turn (Castiglione-Morelli *et al.*, 1995), suggesting that intermediates could perhaps be selected against because of differences in the protein folding.

Such length polymorphism in a coding region of a clock gene raised the question of whether this variation had adaptive value. Are the different length variants equal in the eyes of natural selection? The answer is "probably not" and comes from two lines of evidence, population genetics and behavior. Surveys of 18 populations across Europe and North Africa revealed latitudinal clines in the frequencies of the two most common Thr-Gly variants. While the length variant coding 17 repeats, $(Thr-Gly)_{17}$, is more common in the Mediterranean area and decreases in frequency toward higher latitudes, the $(Thr-Gly)_{20}$ is the most frequent one in northern Europe and shows lower frequencies in the south (Costa _et al._, 1992) (Figure 4.2). Further evidence for the selective nature of the Thr-Gly polymorphism came from the analysis of the nonrandom associations (linkage disequilibrium) between the length alleles and nucleotide polymorphisms in its flanking sequences (Rosato _et al._, 1997a), which showed patterns that are consistent, according to theoretical models (Klitz and Thomson, 1987; Thomson and Klitz, 1987; Robinson _et al._, 1991a,b), with natural selection favoring particular haplotypes.

Although latitudinal clines in genetic polymorphisms can have nonselective explanations such as a combination of founder events and migration, they very often indicate the action of natural selection. Because _per_ controls biological rhythms, two factors that vary with latitude come immediately to mind: photoperiod and temperature. From the patterns of geographical variation in frequencies, it is not possible to choose between the two alternatives (however, see Figure 4.2). One of the most important properties of a biological clock is its ability to maintain an approximately constant period at different temperatures. Ewer _et al._ (1990) showed that arrhythmic per^{01} mutants transformed with a _per_ gene fragment, in which the whole repeat region has been deleted, present a larger variation in the period of its activity rhythms in different temperatures than per^{01} flies transformed with a normal copy of the gene. This finding stimulated Sawyer _et al._ (1997) to examine the temperature compensation of natural-length Thr-Gly variants. The results revealed that the circadian clocks of different Thr-Gly alleles vary in their capacity to maintain a constant period between the temperatures of 18°C and 29°C. These results were backed up by experiments with transgenic per^{01} flies transformed with engineered _per_ genes differing only in the number of Thr-Gly repeats, therefore excluding the possibility that the temperature compensation phenotypes were due to flanking variation that could have been in linkage disequilibrium with the Thr-Gly length variants.

The variant with 20 repeats is the best compensated, while the $(Thr-Gly)_{17}$ variant presents periods of activity closer to 24 hr in high temperatures, suggesting a simple adaptive mechanism by which the former might have additional fitness value in the more extreme temperatures of Northern Europe, while the latter would be better adapted to the warmer and less thermally variable Mediterranean area (Sawyer _et al._, 1997). Although it is unlikely that any one simple explanation can fully account for the Thr-Gly cline, the results are fully consistent with the thermal hypothesis.

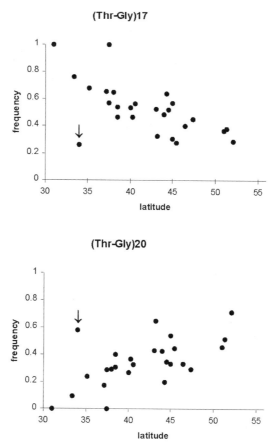

Figure 4.2. Latitudinal clines of *D. melanogaster* Thr-Gly length variants in Europe and North Africa. Graphs on the top and bottom show, respectively, the frequencies of the (Thr-Gly)$_{17}$ and (Thr-Gly)$_{20}$ length alleles against latitude for 26 natural populations (data from Costa *et al.*, 1992, and Rosato *et al.*, 1996). The correlations between latitude and frequency are significant in both cases [Spearman rank correlations: (Thr-Gly)$_{17}$, $R = -0.655$, $p = 0.0003$; (Thr-Gly)$_{20}$, $R = 0.558$, $p = 0.0013$]. The arrows highlight a population (Platanistassa, Cyprus) that is a clear outlier in both graphs, showing a high frequency of (Thr-Gly)$_{20}$ and low frequency of (Thr-Gly)$_{17}$ in a low latitude locality. Interestingly, this is the only high-altitude collection site (1353 m) among the 26 populations. All the other ones are 500 m or below. This suggests that altitudinal clines that parallel the latitudinal variation might exist for the Thr-Gly polymorphism of *D. melanogaster*.

The finding of a clinal variation in the Thr-Gly length polymorphism in *D. melanogaster* raised the question of which kind of geographical patterns one would find in its sibling species, *D. simulans*. As often reported for other genes, such as *G6pd* (Eanes *et al.*, 1993), *prune* (Simmons *et al.*, 1994), hexokinases (Duvernell and Eanes, 2000), and others (Aquadro, 1992; Moriyama and Powell,

1996; Andofatto, 2001), *D. simulans* shows a different pattern of polymorphism in the Thr-Gly region of *per* (Rosato *et al.*, 1994) than its sibling species. There is far less length variation, but, contrary to what is found in *D. melanogaster*, the sequences immediately downstream of the repeats are very variable. Moreover, no significant geographical differentiation among European populations was detected (Rosato *et al.*, 1994). However, when the Tajima test (1989), which compares two measures of genetic variation, one based on the number of segregating sites and the other based on the number of pairwise nucleotide differences, was applied to the data, significant departures from a neutral model were found, indicating that the polymorphism in *D. simulans* is probably under similar balancing selection across different populations in Europe (Rosato *et al.*, 1994).

Comparison of the Thr-Gly repeat region among the species of the *melanogaster* subgroup showed how the repetitive region of the PER protein is evolving by a combination of slippage-like events and point mutations (Peixoto *et al.*, 1992). This process in the long run results in an enormous variation in both sequence and length observed among *Drosophila* species belonging to different groups (Colot *et al.*, 1988; Peixoto *et al.*, 1993). In *D. pseudoobscura*, for example, the Thr-Gly repeats are largely replaced by a very long five-amino-acid degenerate repeat, i.e., units showing variation to the consensus repeat sequence, that is also polymorphic in length, at least among some laboratory strains (Costa *et al.*, 1991). The conformational structure of this pentapeptide generates a β-turn, suggesting that is the functional equivalent to (Thr-Gly)$_3$ (Guantieri *et al.*, 1999).

Interestingly, the repetitive region is very short and conserved in non-drosophilid Diptera (Nielsen *et al.*, 1994; Piccin *et al.*, 2000; Warman *et al.*, 2000; Mazzoni *et al.*, submitted). In fact, within any one *Drosophila* species, variability in the length also correlates with the overall length of the Thr-Gly region, so that the variation observed in the *melanogaster* subgroup, which has a long repetitive region, is high (Peixoto *et al.*, 1992), whereas there is no length variation in the short Thr-Gly region of the *nasuta* subgroup (Zheng *et al.*, 1999). Thus the expansion of the Thr-Gly region observed in some *Drosophila* species seems to represent an evolutionary novelty.

The patterns of length variation observed in the Thr-Gly region of *Drosophila* and other Diptera fit the predictions of models of evolution of repetitive sequences (e.g., Gray and Jeffreys, 1991; Harding *et al.*, 1992). These models predict that only a small proportion of lineages will present large number of repeats and that these lineages will be more polymorphic. However, because the Thr-Gly region represents coding sequences, we would expect constraints on the PER protein, so that only a certain amount of Thr-Gly variation within any one species could be tolerated. In addition, the evolution of large changes in length of the repetitive region between species, for example, *D. melanogaster* and *D. pseudoobscura*, might also be expected to have repercussions for PER protein structure.

We have previously noted that the repetitive regions of the different species have similar predicted secondary structures despite their sequence differences. Thus the large repeat region of D. *pseudoobscura* (> 200 residues) actually has about 30–35 β-turns, whereas that of D. *melanogaster* ($\sim$60 residues) has about 6–8 of these motifs (Costa *et al.*, 1991; Peixoto *et al.*, 1993; Castiglione-Morelli *et al.*, 1995; Guantiere *et al.*, 1999). So what might be the effects of such a large expansion in repeat number on other regions of the protein?

The first clue came from the observation that there is a significant correlation between the interspecific length differences in the repeat region and the amino acid divergence in the flanking sequences when pairs of species were compared (Peixoto *et al.*, 1993). This correlation cannot be accounted for simply by the separation time of the different species, as no significant correlation was found between the intracodon third bases or synonymous changes and the length variation. Therefore, it would seem that the repetitive region and its flanking sequences might be coevolving. This correlation was confirmed using another set of species, which revealed that the upstream "coevolving" flanking region was limited to about 60 residues (Nielsen *et al.*, 1994).

Direct experimental support for the coevolution hypothesis came from experiments using interspecific chimeric genes (Peixoto *et al.*, 1998). The *per*^{01} mutant of D. *melanogaster* transformed with chimeras having the first half of the *melanogaster per* gene and the second half of D. *pseudoobscura per* gene (including the repeats) showed striking differences in percentages of rhythmic flies, depending on the exact position of the chimeric junction (Figure 4.3, see color insert). In one of the constructs (*mps3*), the junction was placed upstream of the flanking sequences that seem to be coevolving with the repeats, resulting in very high levels of rescue. However, in a construct within which the junction was placed between the two regions (*mps2*), the rescue was very poor, suggesting that the D. *pseudoobscura* repeat region does not work well with D. *melanogaster* flanking sequences. Furthermore, in a third construct (*mps5*) the junction was such that it divided the flanking sequence in two. Although this construct gave fairly good levels of rescue, temperature compensation of these transgenic flies' locomotor cycle durations was significantly compromised. Therefore, these results once again link the evolution of Thr-Gly repetitive region to the mechanism of temperature compensation of the clock.

One aspect of *per* not yet fully explored is that it seems to control not only the period of the circadian clock but also interspecific differences in the locomotor activity patterns (Petersen *et al.*, 1988). Variation in the peak of locomotor and mating activities could have important implications for adaptation to different habitats and to temporal reproductive isolation between closely related species (Sakai and Ishida, 2001). Finally, evolutionary studies of *per* were also carried out in relation to its courtship song phenotype and its potential role on speciation, which will be discussed later.

B. Other clock genes

Far fewer evolutionary studies have been carried out with other genes involved in the control of biological rhythms in *Drosophila*. Some of the clock genes cloned to date, such as *double-time* (Kloss *et al.*, 1998), *cryptochrome* (Emery *et al.*, 1998; Stanewsky *et al.*, 1998), and *cycle* (Rutila *et al.*, 1998), seem to be quite conserved when compared with their mammalian homologs. For example, while PER shows only 44% amino acid similarity to its human homolog RIGUI (Sun *et al.*, 1997), CYCLE shows 68% similarity to BMAL1 (Rutila *et al.*, 1998). These more conserved clock genes are unlikely to yield many surprises once sequences from different *Drosophila* species are available, although expression studies and sequence comparisons among more distantly related insects could be very interesting, not only because a higher degree of divergence is expected but also because they could reveal variation in temporal and spatial gene regulation, perhaps reflecting differences in promotor and enhancer sequences.

One *D. melanogaster* gene that is begging for a population-genetic analysis is *Clock* (Allada *et al.*, 1998), because it has a large poly-Q track that has a high chance of being polymorphic in natural populations. If that turns out to be the case, it will be interesting to see if there are any latitudinal clines in the length of the poly-Q that could parallel the one existing in the Thr-Gly region of *per*.

The only other clock gene apart from *per* that has been subject to evolutionary analyses is *timeless* (*tim*), the second *Drosophila* clock gene to be cloned (Myers *et al.*, 1995). Comparison of the TIM protein sequences among *D. virilis*, *D. hydei*, and *D. melanogaster* (Myers *et al.*, 1997; Ousley *et al.*, 1998) revealed that although these sequences are more conserved than PER, there are regions that present a high level of divergence, including the functionally important cytoplasmatic localization domain (CLD). However, PER and TIM show similar levels of divergence between *D. virilis* and *D. melanogaster* in the regions where the two proteins interact, giving some support for the idea of coevolution between the two genes, as proposed by Piccin *et al.* (2000).

One curious feature of *timeless* is the discovery by Rosato *et al.* (1997b) and Myers *et al.* (1997) of a polymorphism in the presence or absence of an alternative upstream methionine initiation site in *D. melanogaster*, causing a 23 amino acid difference in the length of the protein. Although it is not yet clear if this alternative site is actually being used during translation because it seems absent in other species of the *melanogaster* subgroup (Rosato *et al.*, 1997b), there is already preliminary evidence that a latitudinal cline exists in Europe for the relative frequencies of the two forms (M. Zordan, personal communication). It is therefore tempting to speculate whether there might be an interaction for fitness between the *per* and *tim* polymorphisms and what would be the phenotypic effects of the different *per* and *tim* allele combinations.

III. GENES INVOLVED IN SEXUAL BEHAVIOR

A. Male accessory-gland proteins

In the fruit fly, the proteins and peptides that exist in the male seminal fluid substantially influence female behavior after copulation. After mating, a *D. melanogaster* female becomes less attractive, moves less in response to males, and reduces her receptivity to further mating (Wolfner, 1997). The important role male accessory-gland proteins have on female behavior is found not only for *Drosophila*, but also in other insects such as mosquitoes (Lee and Klowden, 1999; Klowden, 1999) and grasshoppers (Hartmann and Loher, 1996).

The male accessory-gland proteins, also known as ACPs, have been studied not only for their behavioral and physiological effects, but also because they might be involved in sexual selection and speciation (Chen, 1996). When a female is mated to two males, the sperm of the second male usually fertilizes most of the eggs. This "sperm displacement" (also known as "sperm precedence") is an important mechanism for sexual selection, and there is evidence that variation in the ability of the first male's sperm to resist the displacement by that of the second male is associated with differences in ACP genes (Clark *et al.*, 1995). Experiments aimed at studying the species-specific effects of the ACPs suggest that these proteins are highly diverged among species (Chen, 1996). There is evidence that differences in the seminal products are likely to be responsible for the reduced insemination observed in some interspecific hybridizations in *Drosophila* (Alipaz *et al.*, 2001) and for conspecific sperm precedence (Price, 1997), i.e., the fact that when a female is mated to two males, one of the same species and one of a sibling species, the majority of the eggs are fertilized by the conspecific sperm, irrespective of the mating order.

The male accessory glands have an estimated number of 80 or more different ACPs (Swanson *et al.*, 2001), ranging from single peptides with fewer than 100 amino acids, which represent about 75% of the products, to prohormone-like and other larger proteins, including one large glycoprotein (Wolfner, 1997). The two best-studied ACPs at both functional and evolutionary levels are the sex-peptide (Acp70A) and the prohormone-like protein Acp26Aa (see later discussion).

The sex peptide Acp70A stimulates egg-laying and induces the females to become nonreceptive to males after mating (Chen *et al.*, 1988). In *D. melanogaster*, it is synthesized as a 55 amino-acid precursor with a 19-residue signal peptide at the N terminus, while the sex peptide itself, i.e., after signal peptide cleavage, has only 36 amino acids (Chen *et al.*, 1988). Comparison of the amino acid sequences of Acp70A in different *Drosophila* species (Figure 4.4) reveals a conserved block at its C-terminal and more divergent N-terminal sequences (Cirera and Aguadé, 1997, 1998a; Chen and Balmer, 1989; Imamura *et al.*, 1998; Schmidt *et al.*, 1993). Cirera and Aguadé (1997) studied the population genetics of the *Acp70A* gene region in

```
D. melanogaster-S    MKTLALFLVLVCVLGLVQSWEWPWNR--KP---TKFPIPSPNPRDKWCRLNLGPAWGGR-C
D. melanogaster-A    MKTLALFLVLVCVLGLVQAWEWPWNR--KP---TKFPIPSPNPRDKWCRLNLGPAWGGR-C
D. simulans          MKTLSIFLVLVCLLGLVQSWEWPWNR--KP---TKFPIPSPNPRDKWCRLNLGPAWGGR-C
D. mauritiana        MKTLSIFLVLVCLLGLVQSWEWPWNR--KP---TKYPIPSPNPRDKWCRLNLGPAWGGR-C
D. sechellia         MKTLSVFLVLVCLLGLVQSWEWPWNR--QP---TRYPIPSPNPRDKWCRLNLGPAWGGR-C
D. suzukii           MKALTLILVLVCIVGLVNSWEWPWNKQKKPWERPRFPIPNPNPRDKWCRLNLGPAWGGR-C
D. subobscura1       MMVPISIMLLLLLVGVALGMPNPMPA--RK-----SSTWGPRDIQKWCRLNFGPAWGGRAC
D. madeirensis1      MMVPISIMLLLLLVGVALGMPNPMPA--RK-----SSTWGPRDIQKWCRLNFGPAWGGRGC
D. subobscura2       MRVPISIMLFLLLLVGVACGVHWRITR--RTT---TSSTWGPRDIQKWCRLNFGPAWGGRGC
D. madeirensis2      MKFPISIMLLLLLVGVALGVHWRITR--RTS---TSSTWGPRDIQKWCRLNFGPAWGGRGC
                     *   :::::  ::*:. .       :      . .*.  :******:******* *
```

Figure 4.4. Protein alignment of the sex peptide (Acp70A) from different *Drosophila* species using ClustalX software (Thompson *et al.*, 1997). It includes two different alleles of *D. melanogaster* (Cirera and Aguadé, 1997) and the duplicated genes of *D. subobscura* and *D. madeirensis* (Cirera and Aguadé, 1998a,b). The first 19 amino acids form the signal peptide (see text). Note the conservation of the C-terminal region. Accession numbers: *D. melanogaster-S* (X99415), *D. melanogaster-A* (X99416), *D. simulans* (X99417), *D. mauritiana* (X99412), *D. sechellia* (X99414), *D. suzukii* (S64573), *D. subobscura1* and *D. subobscura2* (AJ225041), *D. madeirensis1* and *D. madeirensis2* (AJ225032).

D. melanogaster and found an amino acid polymorphism in the signal peptide that could potentially affect the processing of the precursor. The data suggest that one of these alleles is a recent mutation that increased in the population because of natural selection. In addition, Cirera and Aguadé (1997) found that the pattern of nucleotide variation in the 5′ region of the gene, where two highly differentiated haplotypes were found, did not follow the expectations of a neutral model, suggesting that natural selection might be acting on these upstream sequences. This could indicate the presence of yet unidentified regulatory sequences in the region.

In another study, the same authors studied the molecular evolution of a duplication event involving Acp70A in *D. subobscura* and *D. madeirensis* (Cirera and Aguadé, 1998a,b). This duplication appears to be a relatively recent event, around 3 million years ago (MYA), and both copies are probably functional but have diverged considerably (Cirera and Aguadé, 1998a,b) (Figures 4.5A and 4.5B). Because the C-terminal region of the peptide has remained conserved, it would seem that this divergence is not simply a reflection of the relaxation of selective pressure due to the redundancy caused by the presence of an additional copy of the gene. In fact, statistical tests applied to the data indicate significant deviations from neutral expectations, suggesting that natural selection has driven changes in the N-terminal region of the peptide after the duplication event (Cirera and Aguadé, 1998b).

Acp26Aa is a prohormone-like protein sharing sequence homology to the egg-laying hormone of *Aplysia* (Monsma and Wolfner, 1988; Wolfner, 1997). Analysis of mutants indicates that Acp26Aa stimulates egg laying, through oocytes release by the ovary, on the first day of mating, but does not produce changes in

A

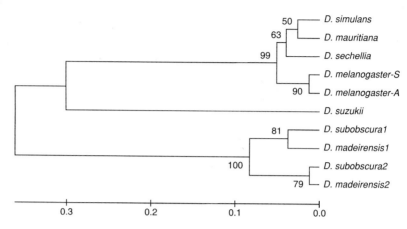

B

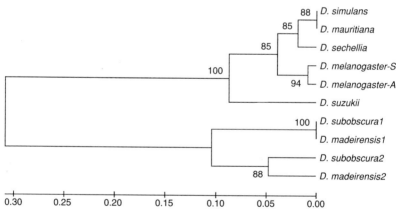

Figure 4.5. (A) Linearized neighbor-joining tree (Saitou and Nei, 1987) of the genes coding the peptides shown in Figure 4.4. Note that the divergence between *D. susukii* and the species of the *D. melanogaster* subgroup is much larger than the divergence between the duplicated genes of *D. subobscura* and *D. madeirensis*. This tree is based only on synonymous sites, and it is probably roughly proportional to the time of divergence of the different genes. (B) Linearized neighbor-joining tree using the peptide sequences shown in Figure 4.4. Note how the divergence between the duplicated genes of *D. subobscura* and *D. madeirensis* is now similar to the divergence of *D. susukii* and the species of the *D. melanogaster* subgroup. This tree reflects the accelerated rate of evolution of the genes of *D. subobscura* and *D. madeirensis* following the duplication event. Both trees were constructed using the molecular evolutionary genetics analysis (MEGA 2.1) software (Kumar *et al.*, 2001). Numbers on the branches represent bootstrap percentage values based on a 1000 replications (Felsenstein, 1985).

female receptivity (to remating) after copulation (Herndon and Wolfner, 1995; Heifetz *et al.*, 2000). The protein has glycosylation and peptidase sites and is processed after transfer to the female; but interestingly enough, this processing requires transfer (during mating) of additional male products, as it does not occur in unmated transgenic females expressing *Acp26Aa* ectopically (Park and Wolfner, 1995).

The *Acp26Aa* gene is located in the left arm of the second chromosome adjacent to another accessory gland protein gene, *Acp26Ab*, and both genes occur within a 2-kb fragment. Aguadé *et al.* (1992) carried out the first evolutionary analysis of the *Acp26A* region by studying polymorphism and divergence in one North American population of *D. melanogaster* and its closely related species, *D. simulans*, *D. mauritiana*, and *D. sechellia*. The data revealed unusually high levels of nonsynonymous variation in the *Acp26Aa* gene of *D. melanogaster*, which was comparable to the level of silent polymorphism, indicating either a lack of constraint or some sort of diversifying selection. Unusually high levels of divergence were also found in the same region when comparisons were made between *D. melanogaster* and its relatives.

In an extension of this early work, Aguadé (1998) obtained 52 more sequences of the *Acp26A* region, including *Acp26Aa* and *Acp26Ab*, from *D. melanogaster* from Europe and Africa. As before, a high level of nonsynonymous variation was detected, but interestingly, no amino acid polymorphism was found in the glycosylation and peptidase sites and in the region showing similarity to the egg-laying hormone of *Aplysia*. Evidence for directional selection, in the form of an excess of fixed nonsynonymous differences between species, was obtained for *Acp26Aa* but not for *Acp26Ab*. These results were corroborated by the findings of Tsaur and Wu (1997) and Tsaur *et al.* (1998). In addition, Tsaur *et al.* (2001) showed that, in *D. mauritiana*, the level of polymorphism in *Acp26Aa* is even higher than its close relative, *D. melanogaster*. The data also suggest that the N- and C-terminal domains of the protein are under different selective pressures. Although both regions show evidence for positive Darwinian selection fixing new favorable mutations, the N terminus, which is cleaved after insemination (Park and Wolfner, 1995), seems to be also under some sort of diversifying selection as it is far more polymorphic than the C-terminal region (Tsaur *et al.*, 2001). The authors suggest that this could be the result of some sort of frequency-dependent sexual selection.

As discussed earlier in the cases of *Acp70A* and *Acp26Aa*, the evolutionary analysis of genes coding for several other accessory-gland proteins (Aguadé, 1999; Begun *et al.*, 2000; Swanson *et al.*, 2001) has indicated that in general they show higher amino acid polymorphism and between-species divergence than most other genes in *Drosophila*. This is likely to be a result of sexual selection acting on these loci as a consequence of cryptic female choice of male seminal products (Eberhard and Cordero, 1995). ACPs are also probably involved in an evolutionary arms race between males and females (Rice, 1996; Holland and Rice, 1999; Clark *et al.*, 1999) because of their detrimental effects on females (Chapman *et al.*, 1995). There is, for example, a significant positive correlation between the female mortality rate after mating and the ability of their mates to resist sperm displacement by a second male (Civetta and Clark, 2000). However, clear examples of the mechanisms by which selection is acting are still missing. Evidence for an association between variation in sperm displacement and some ACP genes has already been reported (Clark *et al.*, 1995). At least in one case, Acp36DE—a large glycoprotein involved in sperm storage (Neubaum and Wolfner, 1999)— this evidence was corroborated by the analysis of mutants in sperm competition experiments (Chapman *et al.*, 2000). More work using a combination of functional and evolutionary analyses will help determining the modes of selection operating on the accessory-gland proteins.

B. Courtship song genes

In more than 100 *Drosophila* species, males vibrate their wings during courtship, producing a kind of love song that seems to increase the receptivity of females (Ewing, 1989; Hall, 1994a; Greenspan and Ferveur, 2000). There is extensive interspecific variation in this acoustic signal (e.g., Cowling and Burnet, 1981; Hoikkala and Lumme, 1987; Wheeler *et al.*, 1988; Ritchie and Gleason, 1995; Noor and Aquadro, 1998), and there is very strong evidence that these song differences are important in the reproductive isolation between closely related species and, as a consequence, in the process of speciation (Kyriacou and Hall, 1982, 1986; Ritchie *et al.*, 1999; Tomaru *et al.*, 1998, 2000; Doi *et al.*, 2001). Moreover, there is also evidence for sexual selection on the intraspecific variation in song parameters (Ritchie *et al.*, 1998; Hoikkala *et al.*, 1998; Hoikkala and Suvanto, 1999). Therefore, song genes are prime suspects for acting as speciation factors (Coyne, 1992).

A few *Drosophila* loci controlling the courtship song have been identified and their genes cloned (Hall, 1994a; Yamamoto *et al.*, 1997; Peixoto and Hall, 1998; Greenspan and Ferveur, 2000). Some of these have been subjected to evolutionary studies as is discussed in the following subsections.

1. period

As mentioned before, the *period* gene was identified by Konopka and Benzer (1971) when they isolated mutants that showed altered circadian rhythms. About 10 years later, Kyriacou and Hall (1980) showed that *period* also controls rhythms in a very different time domain.

The love song that *D. melanogaster* males produce during courtship is composed of a sine song with a carrier frequency of about 160 Hz and a pulse song with an interpulse interval (IPI) around 35 ms (Schilcher, 1976a; Wheeler *et al.*, 1988, 1989). Kyriacou and Hall (1980) discovered that the mean IPI oscillates with a period of about 60 s and that the *per* mutations would accelerate ($per^S = 40$ sec), obliterate (per^O), or slow down ($per^L = 80$ sec) these song rhythms. Moreover, they also showed that in *D. melanogaster*'s sibling species, *D. simulans*, the mean IPI oscillates with a different rhythm (40 sec). Later, using playback experiments, the same authors (Kyriacou and Hall, 1982, 1986) demonstrated that it is the right combination of species-specific mean IPI and song rhythm that is most effective in stimulating females to copulate, indicating an important role that the song differences might play in sexual selection and reproductive isolation. Kyriacou and Hall's findings on the existence of IPI oscillations and their influence on mating-initiation latency were independently confirmed later (Alt *et al.*, 1998; Ritchie *et al.*, 1999), and evidence for rhythms in the love songs have also been found in other *Drosophila* species (Demetriades *et al.*, 1999; Noor and Aquadro, 1998).

Transformation experiments using *D. melanogaster/D. simulans* gene chimeras (Wheeler *et al.*, 1991) not only proved that *per* was responsible for the species differences in the song rhythms but also mapped them to amino acid changes downstream of the Thr-Gly repeats. Although the number of repeats may play no role in the song differences between the two species, deletion of all repeat units does have an effect on the period of the IPI cycling (Yu *et al.*, 1987). It is therefore possible that sexual selection on courtship song variation could also play a role in the maintenance of polymorphisms in the Thr-Gly region in *D. melanogaster* and *D. simulans*, as discussed previously. If that is the case, the adaptive value of the different Thr-Gly variants depends on a combination of their circadian and song effects.

Because of its role in the species-specific differences in a feature of the courtship song in *Drosophila*, *per* was called a "speciation gene" (Coyne, 1992) and became an interesting marker for studies in this area of evolutionary genetics. As mentioned earlier, the species of the *melanogaster* subgroup present variation not only in the number of Thr-Gly repeats but also in the amino acid sequence of the downstream flanking region (Thackeray and Kyriacou, 1990; Peixoto *et al.*, 1992). Some of these differences could influence the song-rhythm variations observed between *D. melanogaster* and its distant relatives within the subgroup, such as *D. yakuba* (Demetriades *et al.*, 1998). In fact, these molecular and behavioral differences become ever more relevant after the remarkable recent discovery of a new species belonging the *melanogaster* subgroup, *D. santomea*, endemic to the São Tomé island on the Gulf of Guinea, West Africa (Lachaise *et al.*, 2000). Analysis of the *per* gene and other markers show that *D. santomea* is a sibling of *D. yakuba*. Both species exist on this island; but whereas *D. santomea* is found in submontane rainforest (above 1100 m), *D. yakuba* occurs at lower altitudes. The two species form an altitudinal hybrid zone with a few rare hybrids detected in the wild. Amino-acid differences in PER were detected between *D. santomea* and *D. yakuba* downstream of the Thr-Gly region that could possibly affect the song rhythms (Lachaise *et al.*, 2000). It will be very interesting to examine the love song of *D. santomea*, in particular the IPI cycling, and its role in prezygotic isolation between the two species.

Kliman and Hey (1993) were among the first to analyze molecular variation outside of the Thr-Gly repeat region in *D. melanogaster* and its close relatives. The data suggested *D. simulans* to be a "parent" species of *D. mauritiana* and *D. sechellia*, a result supported by a much larger study involving 14 genes (Kliman *et al.*, 2000). Ford *et al.* (1994) used *per* to study the variability within and among the three semispecies of the *D. athabasca* complex. Later, Ford and Aquadro (1996) analyzed the differentiation among the three semispecies X-linked genes (including *per*), compared with autosomal loci, and suggested that the X chromosome has been subjected to selective sweeps, probably associated with the courtship song differences that exist among the semispecies.

Gleason and Powell (1997) studied the *per* gene in *D. willistoni* and its sibling species. Comparison of the short Thr-Gly repeat region revealed conservation, although polymorphism and interspecific length variation was found in a downstream region that contains from 6 to 18 Gly repeats. A possible relationship between variation in this region and song variations observed within the *willistoni* group (Ritchie and Gleason, 1995) was not investigated, as song cycles have not been looked for in these species.

Using *per* to study the speciation in the *virilis* group, Hilton and Hey (1996) show that *D. novamexicana* might in fact represent two different cryptic species. They also show that the two chromosomal subspecies of *D. americana* (*D. a. americana* and *D. a. texana*) have extensive gene flow between them despite their different karyotypes. *per* was also used in speciation studies in *D. pseudoobscura* and its sibling species, *D. pseudoobscura bogotana*, *D. persimilis*, and *D. miranda* (Wang and Hey, 1996). These analyses indicated that the genetic variability in the *per* gene of *D. p. bogotana*, an isolated subspecies of *D. pseudoobscura* found in Colombia, is significantly lower than expected and is probably the result of selection. Wang and Hey (1996) also found evidence of introgression between *D. pseudoobscura* and *D. persimilis*. An unusual *D. persimilis* allele shows a region of high similarity to *D. pseudoobscura*, suggesting an old introgression event followed by recombination, since the rest of the sequence shows similarity to other *D. persimilis* alleles. Therefore, the authors reasoned that it is unlikely that *per* plays an important role in the isolation between these two species. However, it is interesting to note that the region of similarity to *D. pseudoobscura* did not include the flanking regions of the Thr-Gly like repeats where the *D. melanogaster*/*D. simulans* song rhythm differences reside (see earlier discussion). Therefore, if *per* is responsible for any of the differences observed in the love song of the two species (Noor and Aquadro, 1998), then one could speculate that this *pseudoobscura* fragment survived in *D. persimilis* only because it "recombined out" the song-controlling region.

2. *cacophony* and other ion-channel genes

cacophony (*cac*) is an X-linked gene in *D. melanogaster* coding for an $\alpha 1$ subunit of a voltage-gated calcium channel (Smith *et al.*, 1996, 1998a; Peixoto *et al.*, 1997). Voltage-gated calcium channels are usually formed by five proteins and are involved in a number of important processes (Stea *et al.*, 1995; Davila, 1999; Jeziorski *et al.*, 2000), and there is strong evidence that *cac* has a role in synaptic transmission (Dellinger *et al.*, 2000). The $\alpha 1$ subunits are large proteins (~ 2000 amino acids) forming the calcium-conducting pore. They have four homologous repeats, each one composed of six putative α-helical transmembrane domains (Stea *et al.*, 1995; Davila, 1999; Jeziorski *et al.*, 2000). *cac* is the site of lethal, visual and song mutations (Kulkarni and Hall, 1987; Smith *et al.*, 1996, 1998a). The *cas*^{S} mutation affects the song in a particularly interesting way. The love song is characterized by longer IPIs and by pulses that are anomalously polycyclic and have higher-than-normal amplitude (Schilcher, 1976b, 1977; Kulkarni and Hall, 1987; Peixoto and Hall, 1998; Smith *et al.*, 1998a). Therefore, *cac* controls features of the love song that usually vary between different populations or different *Drosophila* species (e.g., Cowling and Burnet, 1981; Hoikkala and Lumme, 1987; Ritchie and Kyriacou, 1994; Ritchie *et al.*, 1994; Ritchie and Gleason, 1995; Noor and Aquadro, 1998; Colegrave *et al.*, 2000). In fact, some of the song differences between *D. virilis* and *D. littoralis* have been shown to map to a region of the X chromosome that includes *cac* (Paallysaho *et al.*, 2001).

Comparison of the genomic structures of *cac* and a mammalian calcium channel revealed conservation on the majority of intron/exon junctions (Peixoto *et al.*, 1997). Some of the changes in the genomic structure might not be extremely old, as one of the missing *Drosophila* introns is present in sandflies, another dipteran family (Lins *et al.*, in press). *cac* shows alternative splicing and RNA editing of its transcript (Smith *et al.*, 1996, 1998a,b; Peixoto *et al.*, 1997). These two mechanisms, which generate molecular diversity in channels, raise interesting evolutionary questions. For example, what are the roles of the different splice forms, and how do they evolve? There is evidence that, following the duplication events, they underwent selection for different functions (Peixoto *et al.*, 1997). Supporting this view is the fact that one of the *cac*'s visual mutants that has no song defects maps to one of the two alternative exons that are mutually exclusively spliced (Smith *et al.*, 1998a). With respect to RNA editing, no evolutionary study of *cac* has been carried out, but studies of another ion-channel gene affecting the song (*paralytic*) suggest that the editing might constrain intron divergence (Hanrahan *et al.*, 2000; see later discussion). One important aspect is that the editing in different sites is likely to be independent, and from alternative splicing (Hanrahan *et al.*, 2000). Therefore one can imagine the huge number of possible transcripts resulting from a combination of all different editing sites and alternative exons. These issues raise questions about the selective constraints placed on each individual site (Peixoto *et al.*, 1997). Is the editing of a particular site selected for its average effects on all occurring transcripts, or are certain combinations of edited sites selected for in different tissues?

Peixoto *et al.* (2000) carried out a survey of the molecular variation in a 1-kb fragment in the vicinity of *cac*'s exon 16 (Peixoto *et al.*, 1997) in 27 lines derived from a natural population of D. *melanogaster* in Italy and one D. *simulans* strain. This exon codes for the conserved IIS6 transmembrane domain and the loop linking homologous domains II and III, a region quite divergent among the different calcium channel classes (Stea *et al.*, 1995) and known to be functionally important (e.g., Tanabe *et al.*, 1990; Sheng *et al.*, 1994). The data revealed a low level of polymorphism, with only five noncoding sites variable among the D. *melanogaster* lines. No amino acid substitutions were observed between the two species, and no evidence of selection was found when statistical tests were applied to the data. However, a courtship song analysis of the same lines revealed a significant association between pulse amplitude and one of the polymorphic sites of *cac* (Peixoto *et al.*, 2000). As this site is within an intron, it could be associated with some nonsynonymous polymorphism elsewhere in the gene.

Because the cac^S-mutation was shown to cause frequent convulsions and pronounced locomotor defects when exposed to high temperatures, Peixoto and Hall (1998) examined the courtship song of some other temperature-sensitive physiological mutants. This analysis revealed that genes involved in ion-channel function are a potential source of intraspecific and interspecific genetic variation for song parameters (Peixoto and Hall, 1998). For example, an interesting interaction in the song phenotype—enhancement of the IPI defect coupled with partial suppression of the anomalous polycyclic pulses—was found between *cac* and mle^{napts}, a mutation in a RNA helicase involved in the RNA editing process (Reenan *et al.*, 2000) and previously known for its effects on sodium-channel function (Wu and Ganetzky, 1992). This suggests that mle^{napts} could be affecting *cac*'s proper RNA editing or proper splicing as it does with *paralytic* (Reenan *et al.*, 2000).

slowpoke (*slo*), a gene coding a calcium-activated potassium channel (Atkinson *et al.*, 1991), is another previously known locus that can mutate to cause temperature-sensitive phenotypes and was later identified as a courtship song gene. Two mutant alleles of *slo* present severe song abnormalities including an interaction with *cac* (Peixoto and Hall, 1998), and it is possible that the channels encoded by both genes form part of the pacemaker song generator (Smith *et al.*, 1998a). *slo* has not been studied using an evolutionary approach in relation to its song phenotype, but a comparison of the complex *slo* promotor sequences between D. *melanogaster* and D. *hydei* helped to identify a new exon, a new promotor (Bohm *et al.*, 2000), and three regulatory elements important for the expression in adult muscle, which could have implications for song production (Chang *et al.*, 2000).

The *paralytic* (*para*) gene codes for an $\alpha 1$ subunit of a voltage-gated sodium channel (Loughney *et al.*, 1989). Analysis of the courtship song in *para*[ts1] mutant fly revealed a subtle but interesting song defect. The mean IPI and amplitude in *para*[ts1] varies with temperature in a different way than in wild-type flies. Evidence for interspecific differences in the temperature response of song parameters has been reported in *Drosophila* (Ritchie and Gleason, 1995; Byrne, 1999), and it is possible that *para* is one of the genes controlling this feature of the song.

Like *cacophony*, *para* also shows alternative splicing and RNA editing of its transcript (Loughney *et al.*, 1989; Reenan *et al.*, 2000). Comparison of the developmentally regulated patterns of alternative splicing of *para* in *D. melanogaster* and *D. virilis* show that they are very well conserved (Thackeray and Ganetszky, 1995). Interestingly, the alternative spliced exons show a reduced rate of synonymous changes compared to other exons. RNA editing is also conserved between the two species, although one of the sites studied shows editing in *D. melanogaster* but not in *D. virilis* (Reenan *et al.*, 2000; Hanrahan *et al.*, 2000). The RNA editing process involves a formation of a complex RNA secondary structure, and sequence comparisons allowed the identification of editing complementary sequences (ECS) in unusually conserved intronic regions between *D. melanogaster* and *D. virilis* (Reenan *et al.*, 2000; Hanrahan *et al.*, 2000). The population genetics and evolutionary analysis of intron sequences of *para* and other genes showing RNA editing (Reenan, 2001), will be interesting and could reveal new, previously unknown, editing sites.

3. *nonA*

The *dissonance* mutant allele (*nonA*[diss]) of the *no-on-transient-A* gene causes changes in the courtship song, such that the pulses on a given train become progressively more polycyclic (Kulkarni *et al.*, 1988). Other mutant alleles of this gene affect vision and viability (Jones and Rubim, 1990; Rendahl *et al.*, 1992, 1996; Stanewsky *et al.*, 1993, 1996). *nonA* encodes a putative RNA-binding protein that could be involved in regulating the mRNAs that determine these phenotypes, for example, by binding to 3′ UTRs in eye- or song-specific mRNAs; but at present, the biochemical function of NONA is unknown.

The evolutionary analysis of *nonA* has yielded a few surprises. The first was the accidental discovery of a *nonA*-like gene in *D. melanogaster* (Martin *et al.*, 1995). As this *nonA* homolog apparently lacks introns, it is possible that the duplication event that gave rise to *nonA-like* was mRNA mediated. Despite of the overall divergence in the coding region, there are no premature stop codons; and the important RNA recognition motifs (RRM) are quite conserved, suggesting that this gene is functional (Campesan *et al.*, 2001a). Whether *nonA-like* also has a role in the control of the courtship song remains to be seen.

nonA is tightly linked to the lethal locus *l(1)i19e*, and sequencing analysis of its 5′-flanking region strongly suggested that the latter is the *dGpi1* gene encoding a protein required for the biosynthesis of glycosylphosphatidylinositol (Campesan *et al.*, 2001a). *dGpi1* shows unusually high synonymous substitution rates (Campesan *et al.*, 2001a) and, interestingly enough, its coding sequence overlaps with the promotor sequences of *nonA* (Sandrelli *et al.*, 2001). There is evidence that *dGpi1* constrains the evolution of the regulatory sequences of *nonA*. A significant excess of fixed differences between *D. melanogaster* and *D. simulans* was found in the putative binding sites of regulatory proteins upstream of *nonA*, but not in the region overlapping *dGpi1*, suggesting that *dGpi1* constrains *nonA* promoter evolution (Campesan *et al.*, 2001a). This is particularly interesting because there are enhancer regions within the *dGpi1* transcription unit that are required for proper *nonA* tissue specific expression (Sandrelli *et al.*, 2001).

Comparison of the amino acid sequences of *D. melanogaster* and *D. virilis* (Campesan *et al.*, 2001a) revealed conservation in the region containing the RNA-binding domain and in the adjacent charged region where the site of the *nonA*^{diss} mutation is found (Rendahl *et al.*, 1996). As observed with the *per* gene, the most diverged *nonA* regions are the ones containing repetitive sequences. A large number of differences occur in the first half of the coding region where runs of glycines and other repeats are found in the proteins of both species. The *nonA* gene was also sequenced in *D. littoralis*, a close relative of *D. virilis* (Huttunen *et al.*, submitted). The predicted proteins are very conserved, and no evidence for adaptive fixation of amino acid changes between the two species was found. However, the repetitive regions turn out to be polymorphic in length in *D. littoralis*, and it will be interesting to see if this length variation will have any subtle phenotypic song effects—as was observed for the Thr-Gly natural polymorphisms of *per*, which showed small but adaptively important differences in temperature-compensation of circadian phenotypes (Sawyer *et al.*, 1997).

Genetic analysis of song differences between *D. virilis* and *D. littoralis* mapped some of the main effects to a region of the X chromosome that includes *nonA* (Hoikkala *et al.*, 2000; Paallysaho *et al.*, 2001), suggesting that it might carry species-specific information on the love song. More direct evidence for *nonA* as a reservoir for species-specific information comes from the work of Campesan *et al.* (2001b). The song of *D. virilis* shows polycyclic pulses that resemble those found in the end of *nonA*^{diss} pulse trains (Hoikkala and Lumme, 1987). They also show a tendency to increase the number of cycles per pulse toward the end of trains.

When *D. melanogaster* flies lacking *nonA* were transformed with a fragment containing the *D. virilis nonA* homolog, the song produced by these transgenic flies presented some *virilis*-like features such as a shorter IPI and more

D. melanogaster

D. melanogaster nonA⁻ + nonA^vir

D. virilis

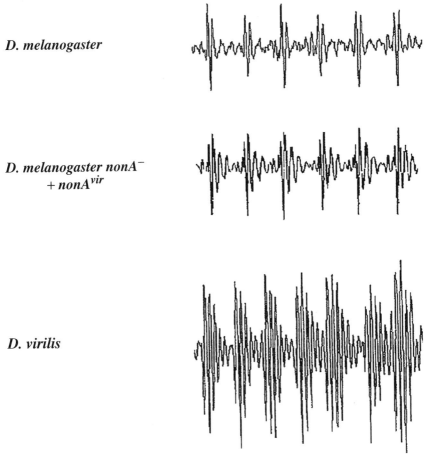

Figure 4.6. Examples of song pulses. (*Top*) *D. melanogaster* wild-type, (*middle*) *D. melanogaster nonA⁻* mutant transformed with the *D. virilis nonA* gene, and (*bottom*) *D. virilis* (courtesy of S. Campesan). Note the polycyclic song of *D. virilis* compared to *D. melanogaster* wild-type and that the transformant carrying a *D. virilis nonA* gene shows a slightly more polycyclic song than wild-type flies. Because the effect is subtle, it was necessary to carry out a sophisticated statistical analysis to confirm the consistency of these song differences. (See Campesan *et al.*, 2001b, for more details.)

polycyclic pulses (Figure 4.6) (Campesan *et al.*, 2001b). Previous work with *nonA* mutations had shown that the visual system was more sensitive to mutation in *nonA*(Rendahl *et al.*, 1992, 1996; Stanewsky *et al.*, 1993, 1996). The *nonA virilis* transformants, however, showed no evidence for visual defects, so the rescue was complete and their song features were not a reflection of a mutant phenotype

(Campesan *et al.*, 2001b). The results were not as dramatic as those obtained for the courtship-song rhythms in the interspecific transformations involving the *per* gene mentioned earlier, probably because the *nonA* study did not involve close relatives but species separated by a 40 to 60 MYA gap (Powell, 1997). Furthermore, the interspecific *per* experiments examined only one specific feature of the song, the period of IPI cycling, and not more general parameters of the song as in the *nonA* study. Nevertheless, the results point to a role for *nonA* in the control of the species-specific differences in the love song (Campesan *et al.*, 2001b).

C. Pheromones

Chemical cues play an important role in sexual behavior and in reproductive isolation for *D. melanogaster* and closely related species (Ferveur *et al.*, 1997; Ferveur, 1997; Savarit *et al.*, 1999). Genetic analysis has indicated that many genes are responsible for the between-species male and female differences in pheromonal cuticular hydrocarbons (Coyne, 1996a,b; Coyne and Charlesworth, 1997). For example, at least six loci on the third chromosome are involved in the *D. sechelia–D. mauritiana* difference in female pheromones (Coyne and Charlesworth, 1997).

Natural populations of *D. melanogaster* harbor a polymorphism in the pheromones produced by females. Most populations around the world produce a hydrocarbon called 7,11-heptacosadiene (7,11-HD), whereas strains from Africa and the Caribbean present low levels of 7,11-HD and high levels of another pheromone, 5,9-heptacosadiene (5,9-HD) (Ferveur *et al.*, 1996). Genetic analysis indicated that this polymorphism was found to be caused by a single factor that mapped to position 87C-D on the third chromosome (Coyne *et al.*, 1999). As it turns out, a pair of desaturase genes (*desat1* and *desat2*) are located within the same region, and the presence or absence of high levels of 5,9-HD was shown to be related to the expression differences of *desat2* (Wicker-Thomas *et al.*, 1997; Dallerac *et al.*, 2000). It was found that—while this gene was actively being expressed in females of a strain producing high levels of 5,9-HD—in females from a 7,11-HD strain, its expression was not detected (Figure 4.7).

These results were further supported by a detailed genetic and DNA sequence analysis of the locus carried out by Takahashi *et al.* (2001). They not only confirmed *desat2* as the locus responsible for the polymorphism, but they also identified a 16-bp deletion in the putative regulatory sequences as the likely cause of the nonexpression observed in 7,11-HD strains. Moreover, when statistical tests were applied to the sequencing data, evidence of natural selection acting on the locus was found. The results suggest that the allele carrying the deletion arose from the undeleted form and was favored by selection during the expansion of *D. melanogaster* out-of-Africa (Takahashi *et al.*, 2001). Therefore, this seems to be a rare example of natural selection favoring a loss-of-function mutation in a gene controlling an aspect of behavior.

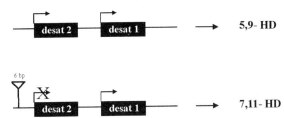

Figure 4.7. Schematic representation of the female cuticular hydrocarbon pheromone polymorphism of *D. melanogaster* controlled by the *desat2* gene. Populations of *D. melanogaster* from Africa and the Caribbean show high levels of the 5,9-HD cuticular hydrocarbon, whereas most populations from other parts of the world show high levels of the 7,11-HD pheromone. This polymorphism is probably caused by a 16-bp deletion in the putative regulatory sequences of *desat2*, inactivating its expression (Dallerac *et al.*, 2000; Takahashi *et al.*, 2001).

IV. CONCLUSION AND PERSPECTIVES

The aim of this review was to give a few examples of studies revolving around the molecular evolution and population genetics of genes controlling behavior in *Drosophila*. All the examples presented here were drawn from circadian rhythms and courtship, mainly because these are the best-studied behaviors at this level. In fact, these two aspects of behavior still have a lot to offer in terms of evolutionary studies. For example, the mapping and future cloning of genes controlling female preference of male love songs (Isoherranen *et al.*, 1999; Doi *et al.*, 2001) will open a new avenue for exploration.

However, many other genes directly or indirectly related to behavior also deserve further attention in the future. For instance, I did not discuss the evolutionary biology of odorant-binding proteins (Hekmat-Scafe *et al.*, 2000). Neither did I discuss the evolution of genes controlling the sex-determination pathway in *Drosophila* (Marin and Baker, 1998; Schutt and Nothiger, 2000). One gene that is particularly interesting in this respect for its behavioral effects is *fruitless* (Ito *et al.*, 1996; Ryner *et al.*, 1996). The evolutionary analysis of this sex-determination gene is still in its early stages (Davis *et al.*, 2000a,b; Gailey *et al.*, 2000), but it will certainly have important implications because of its role in sexual preference, courtship song, and aggression (Villella *et al.*, 1997; Lee and Hall, 2000).

Finally, there are two areas that might turn out to be true goldmines for evolutionary studies related to behavior. One is the study of genes responsible for the development of sexually dimorphic morphological characters that have an effect on mate choice and might be involved in sexual selection and speciation (Kopp *et al.*, 2000). The other area is one that can be called "molecular behavioral ecology." For this, studying the population genetics and evolution of genes controlling behavioral adaptations to the environment, such as foraging strategies (Osborne *et al.*, 1997), might tell us a lot about why animals behave in the way they do in order to survive.

Acknowledgments

I thank Jeff Hall and Bambos Kyriacou for many suggestions about improving the manuscript. Work in my laboratory (which has been reviewed here among discussions of studies performed at many other locations) has been supported by The Wellcome Trust and UNDP/World Bank/WHO Special Programme for Research and Training in Tropical Diseases (TDR), with additional support from Faperj, Fiocruz, and CNPq.

References

Aguadé, M. (1998). Different forces drive the evolution of the *Acp26Aa* and *Acp26Ab* accessory gland genes in the *Drosophila melanogaster* species complex. *Genetics* **150**, 1079–1089.

Aguadé, M. (1999). Positive selection drives the evolution of the Acp29AB accessory gland protein in *Drosophila*. *Genetics* **152**, 543–551.

Aguadé, M., Miyashita, N., and Langley, C. H. (1992). Polymorphism and divergence in the *Mst26A* male accessory gland gene region in *Drosophila*. *Genetics* **132**, 755–770.

Alipaz, J. A., Wu, C.-I., and Karr, T. L. (2001). Gametic incompatibilities between races of *Drosophila melanogaster*. *Proc. R. Soc. Lond. B Biol. Sci.* **268**, 789–795.

Allada, R., White, N. E., So, W. V., Hall, J. C., and Rosbash, M. (1998). A mutant *Drosophila* homologue of mammalian clock disrupts circadian rhythms and transcription of *period* and *timeless*. *Cell* **93**, 791–804.

Alt, S., Ringo, J., Talyn, B., Bray, W., and Dowse, H. (1998). The *period* gene controls courtship song cycles in *Drosophila melanogaster*. *Anim. Behav.* **56**, 87–97.

Andersson, M. B. (1994). "Sexual Selection." Princeton Univ. Press, Princeton.

Andolfatto, P. (2001). Contrasting patterns of X-linked and autosomal nucleotide variation in *Drosophila melanogaster* and *Drosophila simulans*. *Mol. Biol. Evol.* **18**, 279–290.

Aquadro, C. F. (1982). Why is the genome variable? Insights from *Drosophila*. *Trends Genet.* **8**, 355–362.

Atkinson, N. S., Robertson, G. A., and Ganetzky, B. (1991). A component of calcium-activated potassium channels encoded by the *Drosophila-slo* locus. *Science* **253**, 551–555.

Bargiello, T. A., and Young, M. W. (1984). Molecular genetics of a biological clock in *Drosophila*. *Proc. Natl. Acad. Sci. USA* **81**, 2142–2146.

Bargiello, T. A., Jackson, F. R., and Young, M. W. (1984). Restoration of circadian behavioural rhythms by gene transfer in *Drosophila*. *Nature* **312**, 752–754.

Begun, D. J., Whitley, P., Todd, B. L., Waldrip-Dail, H. M., and Clark, A. G. (2000). Molecular population genetics of male accessory gland proteins in *Drosophila*. *Genetics* **156**, 1879–1888.

Bohm, R. A., Wang, B., Brenner, R., and Atkinson, N. S. (2000). Transcriptional control of Ca^{2+}-activated K^+ channel expression: Identification of a second, evolutionarily conserved, neuronal promoter. *J. Exp. Biol.* **203**, 693–704.

Byrne, B. C. (1999). Behaviour-genetic analysis of lovesongs in desert species of *Drosophila*. Ph.D. thesis, University of Leicester, Leicester, U.K.

Campesan, S., Chalmers, D., Sandrelli, F., Megighian, A., Peixoto, A. A., Costa, R., and Kyriacou, C. P. (2001a). Comparative analysis of the *nonA* region in *Drosophila* identifies a highly diverged 5′ gene that may constrain *nonA* promoter evolution. *Genetics* **157**, 751–764.

Campesan, S., Dubrova, Y., Hall, J. C., and Kyriacou, C. K. (2001b). The _nonA_ gene in _Drosophila_ conveys species-specific behavioral characteristics. _Genetics_ **158,** 1535–1543.

Castiglione-Morelli, M. A., Guantieri, V., Villani, V., Kyriacou, C. P., Costa, R., and Tamburro, A. M. (1995). Conformational study of the Thr-Gly repeat in the _Drosophila_ clock protein, PERIOD. _Proc. R. Soc. Lond. B Biol. Sci._ **260,** 155–163.

Ceriani, M. F., Darlington, T. K., Staknis, D., Mas, P., Petti, A. A., Weitz, C. J., and Kay, S. A. (1999). Light-dependent sequestration of TIMELESS by CRYPTOCHROME. _Science_ **295,** 553–556.

Chang, W. M., Bohm, R. A., Strauss, J. C., Kwan, T., Thomas, T., Cowmeadow, R. B., and Atkinson, N. S. (2000). Muscle-specific transcriptional regulation of the _slowpoke_ Ca^{2+}-activated K^{+} channel gene. _J. Biol. Chem._ **275,** 3991–3998.

Chapman, T., Liddle, L. F., Kalb, J. M., Wolfner, M. F., and Partridge, L. (1995). Cost of mating in _Drosophila melanogaster_ females is mediated by male accessory gland products. _Nature_ **373,** 241–244.

Chapman, T., Neubaum, D. M., Wolfner, M. F., and Partridge, L. (2000). The role of male accessory gland protein Acp36DE in sperm competition in _Drosophila melanogaster. Proc. R. Soc. Lond. B Biol. Sci._ **267,** 1097–1105.

Chen, P. S. (1996). The accessory gland proteins in male _Drosophila_: structural, reproductive, and evolutionary aspects. _Experientia_ **52,** 503–510.

Chen, P. S., and Balmer, J. (1989). Secretory proteins and sex peptide of the male accessory gland in _Drosophila sechellia. J. Insect Physiol._ **35,** 759–764.

Chen, P. S., Stumm-Zollinger, E., Aigaki, T., Balmer, J., Bienz, M., and Bohlen, P. (1988). A male accessory gland peptide that regulates reproductive behavior of female _D. melanogaster. Cell_ **54,** 291–298.

Cirera, S., and Aguadé, M. (1997). Evolutionary history of the sex-peptide (_Acp70A_) gene region in _Drosophila melanogaster. Genetics_ **147,** 189–197.

Cirera, S., and Aguadé, M. (1998a). The sex-peptide gene (_Acp70A_) is duplicated in _Drosophila subobscura. Gene_ **210,** 247–254.

Cirera, S., and Aguadé, M. (1998b). Molecular evolution of a duplication: The sex-peptide (_Acp70A_) gene region of _Drosophila subobscura_ and _Drosophila madeirensis. Mol. Biol. Evol._ **15,** 988–996.

Citri, Y., Colot, H. V., Jacquier, A. C., Yu, Q., Hall, J. C., Baltimore, D., and Rosbash, M. (1987). A family of unusually spliced and biologically active transcripts is encoded by a _Drosophila_ clock gene. _Nature_ **326,** 42–47.

Civetta, A., and Clark, A. G. (2000). Correlated effects of sperm competition and postmating female mortality. _Proc. Natl. Acad. Sci. USA_ **97,** 13162–13165.

Clark, A. G., Aguade, M., Prout, T., Harshman, L. G., and Langley, C. H. (1995). Variation in sperm displacement and its association with accessory gland protein loci in _Drosophila melanogaster. Genetics_ **139,** 189–201.

Clark, A. G., Begun, D. J., and Prout, T. (1999). Female × male interactions in _Drosophila_ sperm competition. _Science_ **283,** 217–220.

Colegrave, N., Hollocher, H., Hinton, K., and Ritchie, M. G. (2000). The courtship song of African _Drosophila melanogaster. J. Evol. Biol._ **13,** 143–150.

Colot, H. V., Hall, J. C., and Rosbash, M. (1988). Interspecific comparisons of the _period_ gene of _Drosophila. EMBO J._ **7,** 3929–3937.

Costa, R., Peixoto, A. A., Thackeray, J. R., Dalgleish, R., and Kyriacou, C. P. (1991). Length polymorphism in the Threonine-Glycine-encoding repeat region of the _period_ gene in _Drosophila. J. Mol. Evol._ **32,** 238–246.

Costa, R., Peixoto, A. A., Barbujani, G., and Kyriacou, C. P. (1992). A latitudinal cline in a _Drosophila_ clock gene. _Proc. R. Soc. Lond. B Biol. Sci._ **250,** 43–49.

Cowling, D. E., and Burnet, B. (1981). Courtship songs and genetic control of their acoustic characteristics in sibling species of the _Drosophila melanogaster_ subgroup. _Anim. Behav._ **29,** 924–935.

Coyne, J. A. (1992). Genetics and speciation. _Nature_ **355,** 511–515.

Coyne, J. A. (1996a). Genetics of differences in pheromonal hydrocarbons between *Drosophila melanogaster* and *D. simulans. Genetics* **143,** 353–364.

Coyne, J. A. (1996b). Genetics of a difference in male cuticular hydrocarbons between two sibling species, *Drosophila simulans* and *D. sechellia. Genetics* **143,** 1689–1698.

Coyne, J. A., and Charlesworth, B. (1997). Genetics of a pheromonal difference affecting sexual isolation between *Drosophila mauritiana* and *D. sechellia. Genetics* **145,** 1015–1030.

Coyne, J. A., Wicker-Thomas, C., and Jallon, J. M. (1999). A gene responsible for a cuticular hydrocarbon polymorphism in *Drosophila melanogaster. Genet. Res.* **73,** 189–203.

Curtin, K., Huang, Z. J., and Rosbash, M. (1995). Temporally regulated nuclear entry of the *Drosophila* PERIOD protein contributes to the circadian clock. *Neuron* **14,** 365–372.

Dallerac, R., Labeur, C., Jallon, J. M., Knipple, D. C., Roelofs, W. L., and Wicker-Thomas, C. (2000). A delta 9 desaturase gene with a different substrate specificity is responsible for the cuticular diene hydrocarbon polymorphism in *Drosophila melanogaster. Proc. Natl. Acad. Sci. USA* **97,** 9449–9454.

Darlington, T. K., Wager-Smith, K., Ceriani, M. F, Stankis, D., Gekakis, N., Steeves, T., Weitz, C. J., Takahashi, J., and Kay, S. A. (1998). Closing the circadian loop: CLOCK-induced transcription of its own inhibitors per and tim. *Science* **280,** 1599–1603.

Davila, H. M. (1999). Molecular and functional diversity of voltage-gated calcium channels. *Ann. N.Y. Acad. Sci.* **30,** 131–143.

Davis, T., Kurihara, J., and Yamamoto, D. (2000a). Genomic organisation and characterisation of the neural sex-determination gene *fruitless* (*fru*) in the Hawaiian species *Drosophila heteroneura. Gene* **246,** 143–149.

Davis, T., Kurihara, J., Yoshino, E., and Yamamoto, D. (2000b). Genomic organisation of the neural sex determination gene *fruitless* (*fru*) in the Hawaiian species *Drosophila silvestris* and the conservation of the *fru* BTB protein-protein-binding domain throughout evolution. *Hereditas* **132,** 67–78.

Dellinger, B., Felling, R., and Ordway, R. W. (2000). Genetic modifiers of the *Drosophila* NSF mutant, comatose, include a temperature-sensitive paralytic allele of the calcium channel α1-subunit gene, cacophony. *Genetics* **155,** 203–211.

Demetriades, M. C., Thackeray, J. R., and Kyriacou, C. P. (1998). Courtship song rhythms in *Drosophila yakuba. Anim. Behav.* **57,** 379–386.

Doi, M., Matsuda, M., Tomaru, M., Matsubayashi, H., and Oguma, Y. (2001). A locus for female discrimination behavior causing sexual isolation in *Drosophila. Proc. Natl. Acad. Sci. USA* **98,** 6714–6719.

Dunlap, J. C. (1999). Molecular bases for circadian clocks. *Cell.* **96,** 271–290.

Duvernell, D. D., and Eanes, W. F. (2000). Contrasting molecular population genetics of four hexokinases in *Drosophila melanogaster, D. simulans* and *D. yakuba. Genetics* **156,** 1191–1201.

Eanes, W. F., Kirchner, M., and Yoon, J. (1993). Evidence for adaptive evolution of the G6pd gene in the *Drosophila melanogaster* and *Drosophila simulans* lineages. *Proc. Natl. Acad. Sci. USA* **90,** 7475–7479.

Eberhard, W. G., and Cordero, C. (1995). Sexual selection by cryptic female choice on male seminal products—a new bridge between sexual selection and reproductive physiology. *Trends Ecol. Evol.* **10,** 493–496.

Emery, P., So, W. V., Kaneko, M., Hall, J. C., and Rosbash, M. (1998). CRY, a *Drosophila* clock and light-regulated cryptochrome, is a major contributor to circadian rhythm resetting and photosensitivity. *Cell* **95,** 669–679.

Ewer, J., Hamblen-Coyle, M., Rosbash, M., and Hall, J. C. (1990). Requirement for period gene expression in the adult and not during development for locomotor activity rhythms of imaginal *Drosophila melanogaster. J. Neurogenet.* **7,** 31–73.

Ewing, A. W. (1989). "Arthropod Bioacoustics." Cornell Univ. Press, Ithaca, New York.

Felsenstein, J. (1985). Confidence limits on phylogenies: An approach using the bootstrap. *Evolution* **39,** 783–791.

Ferveur, J. F. (1997). The pheromonal role of cuticular hydrocarbons in *Drosophila melanogaster*. *Bioessays* **19**, 353–358.

Ferveur, J. F., Cobb, M., Boukella, H., and Jallon, J. M. (1996). World-wide variation in *Drosophila melanogaster* sex pheromone: Behavioural effects, genetic bases and potential evolutionary consequences. *Genetica* **97**, 73–80.

Ferveur, J. F., Savarit, F., O'Kane, C. J., Sureau, G., Greenspan, R. J., and Jallon, J. M. (1997). Genetic feminization of pheromones and its behavioral consequences in *Drosophila* males. *Science* **276**, 1555–1558.

Ford, M. J., and Aquadro, C. F. (1996). Selection on X-linked genes during speciation in the *Drosophila athabasca* complex. *Genetics* **144**, 689–703.

Ford, M. J., Yoon, C. K., and Aquadro, C. F. (1994). Molecular evolution of the period gene in *Drosophila athabasca*. *Mol. Biol. Evol.* **11**, 169–182.

Gailey, D. A., Ho, S. K., Ohshima, S., Liu, J. H., Eyassu, M., Washington, M. A., Yamamoto, D., and Davis, T. (2000). A phylogeny of the Drosophilidae using the sex-behaviour gene *fruitless*. *Hereditas* **133**, 81–83.

Gleason, J. M, and Powell, J. R. (1997). Interspecific and intraspecific comparisons of the *period* locus in the *Drosophila willistoni* sibling species. *Mol. Biol. Evol.* **14**, 741–753.

Gotter, A. L., Levine, J. D., and Reppert, S. M. (1999). Sex-linked *period* genes in the silkmoth, *Antheraea pernyi*: implications for circadian clock regulation and the evolution of sex chromosomes. *Neuron* **24**, 953–965.

Gray, I. C., and Jeffreys, A. J. (1991). Evolutionary transience of hypervariable minisatellites in man and the primates. *Proc. R. Soc. Lond. B Biol. Sci.* **243**, 241–253.

Greenspan, R. J., and Ferveur, J. F. (2000). Courtship in *Drosophila*. *Annu. Rev. Genet.* **34**, 205–232.

Guantieri, V., Pepe, A., Zordan, M., Kyriacou, C. P., Costa, R., and Tamburro, A. M. (1999). Different *period* gene repeats take "turns" at fine-tuning the circadian clock. *Proc. R. Soc. Lond. B Biol. Sci.* **266**, 2283–2288.

Hall, J. C. (1994a). The mating of a fly. *Science* **264**, 1702–1714.

Hall, J. C. (1994b). Pleiotropy of behavioral genes. In "Flexibility and Constraint in Behavioral Systems" (R. J. Greenspan and C. P. Kyriacou, eds.), pp. 15–27. John Wiley & Sons, Ltd.

Hall, J. C. (1998). Genetics of biological rhythms in *Drosophila*. *Adv. Genet.* **33**, 135–184.

Hamblen, M. J., White, N. E., Emery, P. T., Kaiser, K., and Hall, J. C. (1998). Molecular and behavioral analysis of four *period* mutants in *Drosophila melanogaster* encompassing extreme short, novel long, and unorthodox arrhythmic types. *Genetics*. **149**, 165–178.

Hanrahan, C. J., Palladino, M. J., Ganetzky, B., and Reenan, R. A. (2000). RNA editing of the *Drosophila para* Na$^{(+)}$ channel transcript. Evolutionary conservation and developmental regulation. *Genetics* **155**, 1149–1160.

Hardin, P. E., Hall, J. C., and Rosbash, M. (1990). Feedback of the *Drosophila period* gene product on circadian cycling of its messenger RNA levels. *Nature* **343**, 536–540.

Harding, R. M., Boyce, A. J., and Clegg, J. B. (1992). The evolution of tandemly repetitive DNA: Recombination rules. *Genetics* **132**, 847–859.

Hartmann, R., and Loher, W. (1996). Control mechanisms of the behavior 'secondary defense' in the grasshopper *Gomphocerus rufus* L. (Gomphocerinae: Orthoptera). *J. Comp. Physiol. A* **178**, 329–336.

Heifetz, Y., Lung, O., Frongillo, E. A. Jr, and Wolfner, M. F. (2000). The *Drosophila* seminal fluid protein Acp26Aa stimulates release of oocytes by the ovary. *Curr. Biol.* **10**, 99–102.

Hekmat-Scafe, D. S., Dorit, R. L., and Carlson, J. R. (2000). Molecular evolution of odorant-binding protein genes OS-E and OS-F in *Drosophila*. *Genetics* **155**, 117–127.

Herndon, L. A., and Wolfner, M. F. (1995). A *Drosophila* seminal fluid protein, Acp26Aa, stimulates egg laying in females for 1 day after mating. *Proc. Natl. Acad. Sci. USA* **92**, 10114–10118.

Hilton, H., and Hey, J. (1996). DNA sequence variation at the period locus reveals the history of species and speciation events in the Drosophila virilis group. Genetics **144**, 1015–1025.

Hoikkala, A., and Lumme, J. (1987). The genetic basis of evolution of the male courtship sounds in the Drosophila virilis group. Evolution **41**, 827–845.

Hoikkala, A., and Suvanto, L. (1999). Male courtship song frequency as an indicator of male mating success in Drosophila montana. J. Insect Behav. **12**, 599–609.

Hoikkala, A., Aspi, J., and Suvanto, L. (1998). Male courtship song frequency as an indicator of male genetic quality in an insect species, Drosophila montana. Proc. R. Soc. Lond. B Biol. **265**, 503–508.

Hoikkala, A., Paallysaho, S., Aspi, J., and Lumme, J. (2000). Localization of genes affecting species differences in male courtship song between Drosophila virilis and D. littoralis. Genet. Res. **75**, 37–45.

Holland, B., and Rice, W. R. (1999). Experimental removal of sexual selection reverses intersexual antagonistic coevolution and removes a reproductive load. Proc. Natl. Acad. Sci USA **96**, 5083–5088.

Huttunen, S., Campesan, S., and Hoikkala, A. Nucleotide variation at the no-on-transient-A gene in Drosophila littoralis. Submitted.

Imamura, M., Haino-Fukushima, K., Aigaki, T., and Fuyama, Y. (1998). Ovulation stimulating substances in Drosophila biarmipes males: their origin, genetic variation in the response of females, and molecular characterization. Insect Biochem. Mol. Biol. **28**, 365–372.

Isoherranen, E., Aspi, J., and Hoikkala, A. (1999). Inheritance of species differences in female receptivity and song requirement between Drosophila virilis and D. montana. Hereditas **131**, 203–209.

Ito, H., Fujitani, K., Usui, K., Shimizu-Nishikawa, K., Tanaka, S., and Yamamoto, D. (1996). Sexual orientation in Drosophila is altered by the satori mutation in the sex-determination gene fruitless that encodes a zinc finger protein with a BTB domain. Proc. Natl. Acad. Sci. USA **93**, 9687–9692.

Jackson, F. R., Bargiello, T. A., Yun, S. H., and Young, M. W. (1986). Product of per locus of Drosophila shares homology with proteoglycans. Nature **320**, 185–188.

Jeziorski, M. C., Greenberg, R. M., and Anderson, P. A. V. (2000). The molecular biology of invertebrate voltage-gated Ca^{2+} channels. J. Exp. Biol. **203**, 841–856.

Jones, K. R., and Rubin, G. M. (1990). Molecular analysis of no-on-transient-A, a gene required for normal vision in Drosophila. Neuron **4**, 711–723.

Kliman, R. M., and Hey, J. (1993). DNA sequence variation at the period locus within and among species of the Drosophila melanogaster complex. Genetics **133**, 375–387.

Kliman, R. M., Andolfatto, P., Coyne, J. A., Depaulis, F., Kreitman, M., Berry, A. J, McCarter, J., Wakeley, J., and Hey, J. (2000). The population genetics of the origin and divergence of the Drosophila simulans complex species. Genetics **156**, 1913–1931.

Klitz, W., and Thomson, G. (1987). Disequilibrium pattern analysis. II. Application to Danish HLA A and B locus data. Genetics **116**, 633–643.

Kloss, B., Price, J. L., Saez, L., Blau, J., Rothenfluh, A., Wesley, C. S., and Young, M. W. (1998). The Drosophila clock gene double-time encodes a protein closely related to human casein kinase 1ε Cell **94**, 97–107.

Klowden, M. J. (1999). The check is in the male: Male mosquitoes affect female physiology and behavior. J. Am. Mosq. Control Assoc. **15**, 213–220.

Konopka, R. J., and Benzer, S. (1971). Clock mutants of Drosophila melanogaster. Proc. Natl. Acad. Sci. USA **68**, 2112–2116.

Kopp, A., Duncan, I., and Carroll, S. B. (2000). Genetic control and evolution of sexually dimorphic characters in Drosophila. Nature **408**, 553–559.

Kulkarni, S. J., and Hall, J. C. (1987). Behavioral and cytogenetic analysis of the cacophony courtship song mutant and interacting genetic variants in Drosophila melanogaster. Genetics **115**, 461–475.

Kulkarni, S. J., Steinlauf, A. F., and Hall, J. C. (1988). The dissonance mutant of Drosophila melanogaster: isolation, behavior and cytogenetics. Genetics **118**, 267–285.

Kumar, S., Tamura K., Jakobsen I., and Nei, M. (2001). MEGA2: Molecular evolutionary genetics analysis software. *Bioinformatics* (in press).

Kyriacou, C. P., and Hall, J. C. (1980). Circadian rhythm mutations in *Drosophila* affect short-term fluctuations in the male's courtship song. *Proc. Natl. Acad. Sci. USA* **77,** 6729–6733.

Kyriacou, C. P., and Hall, J. C. (1982). The function of courtship song rhythms in *Drosophila. Anim. Behav.* **30,** 784–801.

Kyriacou, C. P., and Hall, J. C. (1986). Interspecific genetic control of courtship song production and reception in *Drosophila. Science* **232,** 494–497.

Lachaise, D., Harry, M., Solignac, M., Lemeunier, F., Benassi, V. V., and Cariou, M. L. (2000). Evolutionary novelties in islands: *Drosophila santomea,* a new *melanogaster* sister species from Sao Tome. *Proc. R. Soc. Lond. B Biol. Sci.* **267,** 1487–1495.

Lee, G., and Hall, J. C. (2000). A newly uncovered phenotype associated with the *fruitless* gene of *Drosophila melanogaster*: Aggression-like head interactions between mutant males. *Behav. Genet.* **30,** 263–275.

Lee, J. J., and Klowden, M. J. (1999). A male accessory gland protein that modulates female mosquito (Diptera: Culicidae) host-seeking behavior. *J. Am. Mosq. Control Assoc.* **15,** 4–7.

Lins, R, Oliveira, S. G., Souza, N. A., de Queiroz, R. G., Justiniano, S. C. B, Ward, R. D., Kyriacou, C. P., and Peixoto, A. A. Molecular evolution of the *cacophony* IVS6 region in sandflies. *Ins. Mol. Biol.* (in press).

Loughney, K., Kreber, R., and Ganetzky, B. (1989). Molecular analysis of the *para* locus, a sodium channel gene in *Drosophila. Cell* **58,** 1143–1154.

Marin, I., and Baker, B. S. (1998). The evolutionary dynamics of sex determination. *Science* **281,** 1990–1994.

Martin, C. H, Mayeda, C. A., Davis, C. A., Ericsson, C. L., Knafels, J. D., Mathog, D. R., Celniker, S. E., Lewis, E. B., and Palazzolo, M. J. (1995). Complete sequence of the *bithorax* complex of *Drosophila. Proc. Natl. Acad. Sci. USA* **92,** 8398–8402.

Martinek, S., Inonog, S., Manoukian, A. S., and Young, M. W. (2001). A role for the segment polarity gene *shaggy/GSK-3* in the *Drosophila* circadian clock. *Cell* **105,** 769–779.

Mazzoni, C. J., Gomes, C. A., Souza, N. A., de Queiroz, R. G., Justiniano, S. C. B., Ward, R. D., Kyriacou, C. P., and Peixoto, A. A. Molecular evolution of the *period* gene and phylogeny of Neotropical sandflies. Submitted.

Monsma, S. A., and Wolfner, M. F. (1988). Structure and expression of a *Drosophila* male accessory gland gene whose product resembles a peptide pheromone precursor. *Genes Dev.* **2,** 1063–1073.

Moriyama, E. N., and Powell, J. R. (1996). Intraspecific nuclear DNA variation in *Drosophila. Mol. Biol. Evol.* **13,** 261–277.

Myers, M. P., Wager-Smith, K., Wesley, C. S., Young, M. W., and Sehgal, A. (1995). Positional cloning and sequence analysis of the *Drosophila* clock gene *timeless. Science* **270,** 805–808.

Myers, M. P., Rothenfluh, A., Chang, M., and Young, M. W. (1997). Comparison of chromosomal DNA composing timeless in *Drosophila melanogaster* and *D. virilis* suggests a new conserved structure for the TIMELESS protein. *Nucleic Acids Res.* **25,** 4710–4714.

Neubaum, D. M., and Wolfner, M. F. (1999). Mated *Drosophila melanogaster* females require a seminal fluid protein, Acp36DE, to store sperm efficiently. *Genetics* **153,** 845–857.

Nielsen, J., Peixoto, A. A., Piccin, A., Costa, R., Kyriacou, C. P., and Chalmers, D. (1994). Big flies, small repeats: The "Thr-Gly" region of the *period* gene in Diptera. *Mol. Biol. Evol.* **11,** 839–853.

Noor, M. A. F, and Aquadro, C. F. (1998). Courtship songs of *Drosophila pseudoobscura* and *D. persimilis*: analysis of variation. *Anim. Behav.* **56,** 115–125.

Osborne, K. A., Robichon, A., Burgess, E., Butland, S., Shaw, R. A., Coulthard, A., Pereira, H. S, Greenspan, R. J., and Sokolowski, M. B. (1997). Natural behavior polymorphism due to a cGMP-dependent protein kinase of *Drosophila. Science* **277,** 834–836.

Ousley, A., Zafarullah, K., Chen, Y., Emerson, M., Hickman, L., and Sehgal, A. (1998). Conserved regions of the *timeless* (*tim*) clock gene in *Drosophila* analyzed through phylogenetic and functional studies. *Genetics* **148**, 815–825.

Paallysaho, S., Huttunen, S., and Hoikkala, A. (2001). Identification of X chromosomal restriction fragment length polymorphism markers and their use in a gene localization study in *Drosophila virilis* and *D. littoralis*. *Genome* **44**, 242–248.

Park, M., and Wolfner, M. F. (1995). Male and female cooperate in the prohormone-like processing of a *Drosophila melanogaster* seminal fluid protein. *Dev. Biol.* **171**, 694–702.

Peixoto, A. A., and Hall, J. C. (1998). Analysis of temperature-sensitive mutants reveals new genes involved in the courtship song of *Drosophila*. *Genetics* **148**, 827–838.

Peixoto, A. A., Costa, R., Wheeler, D. A., Hall, J. C., and Kyriacou, C. P. (1992). Evolution of the Threonine-Glycine repeat region of the *period* gene in the *melanogaster* species subgroup of *Drosophila*. *J. Mol. Evol.* **35**, 411–419.

Peixoto, A. A., Campesan, S., Costa, R., and Kyriacou, C. P. (1993). Molecular evolution of a repetitive region within the *per* gene of *Drosophila*. *Mol. Biol. Evol.* **10**, 127–139.

Peixoto, A. A., Smith, L. A., and Hall, J. C. (1997). Genomic organization and evolution of alternative exons in a *Drosophila* calcium channel gene. *Genetics* **145**, 1003–1013.

Peixoto, A. A., Hennessy, M., Townson, I., Hasan, G., Rosbash, M., Costa, R., and Kyriacou, C. P. (1998). Molecular coevolution within a clock gene in *Drosophila*. *Proc. Nat. Acad. Sci. USA* **95**, 4475–4480.

Peixoto, A. A., Costa, R., and Hall, J. C. (2000). Molecular and behavioral analysis of sex-linked courtship song variation in a natural population of *Drosophila melanogaster*. *J. Neurogenet.* **14**, 245–256.

Petersen, G., Hall, J. C., and Rosbash, M. (1988). The *period* gene of *Drosophila* carries species-specific behavioural instructions. *EMBO J.* **7**, 3939–3947.

Piccin, A., Couchman, M., Clayton, J. D., Chalmers, D., Costa, R., and Kyriacou, C. P. (2000). The clock gene *period* of the housefly, *Musca domestica*, rescues behavioral rhythmicity in *Drosophila melanogaster*. Evidence for intermolecular coevolution? *Genetics* **154**, 747–758.

Pittendrigh, C. S. (1993). Temporal organization: reflections of a Darwinian clock-watcher. *Annu. Rev. Physiol.* **55**, 16–54.

Powell, J. R. (1997). "Progress and Prospects in Evolutionary Biology: The *Drosophila* Model." Oxford Univ. Press, Oxford.

Price, C. S. (1997). Conspecific sperm precedence in *Drosophila*. *Nature* **388**, 663–666.

Price, J. L., Blau, J., Rothenfluh, A., Abodeely, M., Kloss, B., and Young, M. W. (1998). *double-time* is a novel *Drosophila* clock gene that regulates PERIOD protein accumulation. *Cell* **94**, 83–95.

Reddy, P., Zehring, W. A., Wheeler, D. A., Pirrota, V., Hadfield, C., Hall, J. C., and Rosbash, M. (1984). Molecular analysis of the *period* locus in *Drosophila melanogaster* and identification of a transcript involved in biological rhythms. *Cell* **38**, 701–710.

Reenan, R. A. (2001). The RNA world meets behavior: A→I pre-mRNA editing in animals. *Trends Genet.* **17**, 53–56.

Reenan, R. A., Hanrahan, C. J., and Barry, G. (2000). The *mle* (*napts*) RNA helicase mutation in *Drosophila* results in a splicing catastrophe of the *para* Na$^+$ channel transcript in a region of RNA editing. *Neuron* **25**, 139–149.

Regier, J. C., Fang, Q. Q., Mitter, C., Peigler, R. S., Friedlander, T. P., and Solis, M. A. (1998). Evolution and phylogenetic utility of the *period* gene in Lepidoptera. *Mol. Biol. Evol.* **15**, 1172–1182.

Rendahl, K. G., Jones, K. R., Kulkarni, S. J., Bagully, S. H., and Hall, J. C. (1992). The *dissonance* mutation of the *no-on-transient-A* locus of *Drosophila melanogaster*: Genetic control of courtship and visual behaviors by a protein with putative RNA-binding motifs. *J. Neurosci.* **12**, 390–407.

Rendahl, K. G., Gaukhshteyn, N., Wheeler, D. A., Fry, T. A., and Hall, J. C. (1996). Defects in courtship and vision caused by amino acid substitutions in a putative RNA-binding protein encoded by the *no-on-transient-A* (*nonA*) gene of *Drosophila*. *J. Neurosci.* **16**, 1511–1522.

Reppert, S. M., Tsai, T., Roca, A. L., and Sauman, I. (1994). Cloning of a structural and functional homolog of the circadian clock gene *period* from the giant silkmoth *Antheraea pernyi*. *Neuron* **13**, 1167–1176.

Rice, W. R. (1996). Sexually antagonistic male adaptation triggered by experimental arrest of female evolution. *Nature* **381**, 232–234.

Ripperger, J. A, and Schibler, U. (2001). Circadian regulation of gene expression in animals. *Curr. Opin. Cell. Biol.* **13**, 357–362.

Ritchie, M. G., and Gleason, J. M. (1995). Rapid evolution of courtship song pattern in *Drosophila willistoni* sibling species. *J. Evol. Biol.* **8**, 463–479.

Ritchie, M. G., and Kyriacou, C. P. (1994). Genetic variability of courtship song in a population of *Drosophila melanogaster*. *Anim. Behav.* **48**, 425–434.

Ritchie, M. G., Yate, V. H., and Kyriacou, C. P. (1994). Genetic variability of the interpulse interval of courtship song among some European populations of *Drosophila melanogaster*. *Heredity* **72**, 459–464.

Ritchie, M. G., Townhill, R. M., and Hoikkala, A. (1998). Female preference for fly song: Playback experiments confirm the targets of sexual selection. *Anim. Behav.* **56**, 713–717.

Ritchie, M. G, Halsey, E. J., and Gleason, J. M. (1999). *Drosophila* song as a species-specific mating signal and the behavioural importance of Kyriacou and Hall cycles in *D. melanogaster* song. *Anim. Behav.* **58**, 649–657.

Robinson, W. P., Asmussen, M. A., and Thomson, G. (1991a). Three-locus systems impose additional constraints on pairwise disequilibria. *Genetics* **129**, 925–930.

Robinson, W. P., Cambon-Thomsen, A., Borot, N., Klitz, W., and Thomson, G. (1991b). Selection, hitchhiking and disequilibrium analysis at three linked loci with application to HLA data. *Genetics* **129**, 931–948.

Rosato, E., Peixoto, A. A., Barbujani, G., Costa, R., and Kyriacou, C. P. (1994). Molecular polymorphism in the *period* gene of *Drosophila simulans*. *Genetics* **138**, 693–707.

Rosato, E., Peixoto, A. A., Gallippi, A., Kyriacou, C. P., and Costa, R. (1996). Mutational mechanisms, phylogeny, and evolution of a repetitive region within a clock gene of *Drosophila melanogaster*. *J. Mol. Evol.* **42**, 392–408.

Rosato, E., Peixoto, A. A., Costa, R., and Kyriacou, C. P. (1997a). Linkage disequilibrium, mutational analysis, natural selection in the repetitive region of the clock gene *period* in *Drosophila melanogaster*. *Genet. Res.* **69**, 89–99.

Rosato, E., Trevisan, A., Sandrelli, F., Zordan, M., Kyriacou, C. P., and Costa, R. (1997b). Conceptual translation of *timeless* reveals alternative initiating methionines in *Drosophila*. *Nucleic Acids Res.* **25**, 455–458.

Rutila, J. E., Vipin, S., Le, M., So, W. V., Rosbash, M., and Hall, J. C. (1998). CYCLE is a second bHLH-PAS Clock protein essential for circadian rhythmicity and transcription of *Drosophila period* and *timeless*. *Cell* **93**, 805–814.

Ryner, L. C., Goodwin, S. F., Castrillon, D. H., Anand, A., Villella, A., Baker, B. S., Hall, J. C., Taylor, B. J., and Wasserman, S. A. (1996). Control of male sexual behavior and sexual orientation in *Drosophila* by the *fruitless* gene. *Cell* **87**, 1079–1089.

Saez, L., and Young, M. W. (1996). Regulation of nuclear entry of the *Drosophila* clock proteins period and timeless. *Neuron* **17**, 911–920.

Saitou, N., and Nei, M. (1987). The neighbor-joining method: A new method for reconstructing phylogenetic trees. *Mol. Biol. Evol.* **4**, 406–425.

Sakai, T., and Ishida, N. (2001). Circadian rhythms of female mating activity governed by clock genes in *Drosophila*. *Proc. Natl. Acad. Sci. USA* **98**, 9221–9225.

Sandrelli, F., Campesan, S., Rossetto, M., Benna, C., Zieger, E., Megighian, A., Couchman, M., Kyriacou, C. P., and Costa, R. (2001). Molecular dissection of the 5′ region of *no-on-transient-A* of *Drosophila melanogaster* reveals cis-regulation by adjacent *dGpi1* sequences. *Genetics* **157**, 765–775.

Sauman, I., and Reppert, S. M. (1996). Circadian clock neurons in the silkmoth *Antheraea pernyi*: Novel mechanisms of Period protein regulation. *Neuron* **17**, 889–900.

Savarit, F., Sureau, G., Cobb, M., and Ferveur, J. F. (1999). Genetic elimination of known pheromones reveals the fundamental chemical bases of mating and isolation in *Drosophila. Proc. Natl. Acad. Sci. USA* **96,** 9015–9020.

Sawyer, L., Hennessy, M., Peixoto, A. A., Rosato, E., Parkinson, H., Costa, R, and Kyriacou, C. P. (1997). Natural variation in a *Drosophila* clock gene and temperature compensation. *Science* **278,** 2117–2120.

Schilcher, F. v. (1976a). The function of sine song and pulse song in the courtship of *Drosophila melanogaster. Anim. Behav.* **24,** 622–625.

Schilcher, F. v. (1976b). The behavior of *cacophony*, a courtship song mutant in *Drosophila melanogaster. Behav. Biol.* **17,** 187–196.

Schilcher, F. v. (1977). A mutant which changes courtship song in *Drosophila melanogaster. Behav. Genet.* **7,** 251–259.

Schmidt, T., Choffat, Y., Schneider, M., Hunziker, P., Fuyama, Y., and Kubli, E. (1993). *Drosophila suzukii* contains a peptide homologous to the *Drosophila melanogaster* sex-peptide and functional in both species. *Insect Biochem. Mol. Biol.* **23,** 571–579.

Schutt, C., and Nothiger, R. (2000). Structure, function, and evolution of sex-determining systems in Dipteran insects. *Development* **127,** 667–677.

Sehgal, A., Rothenflush-Hilfiker, A., Hunter-Ensor, M., Chen, Y., Myers, M. P., and Young, M. W. (1995). Rhythmic expression of *timeless*, a basis for promoting circadian cycles in *period* gene autoregulation. *Science* **270,** 808–810.

Sheng, Z.-H., Rettig, J., Takahashi, M., and Catterall, W. (1994). Identification of a syntaxin-binding site on N-type calcium channels. *Neuron* **13,** 1303–1313.

Simmons, G. M., Kwok, W., Matulonis, P., and Venkatesh, T. (1994). Polymorphism and divergence at the *prune* locus in *Drosophila melanogaster* and *D. simulans. Mol. Biol. Evol.* **11,** 666–671.

Smith, L. A., Wang, X. J., Peixoto, A. A., Neumann, E. K., Hall, L. M., and Hall, J. C. (1996). A *Drosophila* calcium channel α-1 subunit gene maps to a genetic locus associated with behavioral and visual defects. *J. Neurosci.* **16,** 7868–7879.

Smith, L. A., Peixoto, A. A., Kramer, E. M., Villella, A., and Hall, J. C. (1998a). Courtship and visual defects of *cacophony* mutants reveal functional complexity of a calcium-channel α-1 subunit in *Drosophila. Genetics* **149,** 1407–1426.

Smith, L. A., Peixoto, A. A., and Hall, J. C. (1998b). RNA editing in the *Drosophila* DMCA1A calcium-channel α-1 subunit transcript. *J. Neurogenet.* **12,** 227–240.

Stanewsky, R., Rendahl, K. G., Dill, M., and Saumweber, H. (1993). Genetic and molecular analysis of the X chromosomal region 14B17–14C4 in *Drosophila melanogaster*: Loss of function in NONA, a nuclear protein common to many cell types, results in specific physiological and behavioral defects. *Genetics* **135,** 419–442.

Stanewsky, R., Fry, T. A., Reim, I., Saumweber, H., and Hall, J. C. (1996). Bioassaying putative RNA-binding motifs in a protein encoded by a gene that influences courtship and visually mediated behavior in *Drosophila: In vitro* mutagenesis of *nonA. Genetics* **143,** 259–275.

Stanewsky, R., Kaneko, M., Emery, P., Beretta, B., Wager-Smith, K., Kay, S., Rosbash, M., and Hall, J. C. (1998). The *cryb* mutation identifies *cryptochrome* as a circadian photoreceptor in *Drosophila. Cell* **95,** 681–692.

Stea, A., Soong, T. W., and Snutch, T. P. (1995). Voltage-gated calcium channels. *In* "Ligand- and Voltage-Gated Ion Channels" (R. A. North, ed.), pp. 114–151. CRC Press, Inc., Boca Raton, FL.

Sun, Z. S., Albrecht, U., Zhuchenko, O., Bailey, J., Eichele, G., and Lee, C. C. (1997). RIGUI, a putative mammalian orthologue of the *Drosophila period* gene. *Cell* **90,** 1003–1011.

Swanson, W. J., Clark, A. G., Waldrip-Dail, H. M., Wolfner, M. F., and Aquadro, C. F. (2001). Evolutionary EST analysis identifies rapidly evolving male reproductive proteins in *Drosophila. Proc. Natl. Acad. Sci. USA* **98,** 7375–7379.

Tajima, F. (1989). Statistical method for testing the neutral mutation hypothesis by DNA polymorphism. *Genetics* **123,** 585–595.

Takahashi, A., Tsaur, S. C., Coyne, J. A., and Wu, C. I. (2001). The nucleotide changes governing cuticular hydrocarbon variation and their evolution in _Drosophila melanogaster_. _Proc. Natl. Acad. Sci. USA_ **98**, 3920–3925.

Tanabe, T., Beam, K. G., Adams, B. A., Niidome, T., and Numa, S. (1990). Regions of the skeletal muscle dihydropyridine receptor critical for excitation–contraction coupling. _Nature_ **346**, 567–569.

Thackeray, J. R., and Ganetzky, B. (1995). Conserved alternative splicing patterns and splicing signals in the _Drosophila_ sodium channel gene _para_. _Genetics_ **141**, 203–214.

Thackeray, J. R., and Kyriacou, C. P. (1990). Molecular evolution in the _Drosophila yakuba_ period locus. _J. Mol. Evol._ **31**, 389–401.

Thompson, J. D., Gibson, T. J., Plewniak, F., Jeanmougin, F., and Higgins, D. G. (1997). The ClustalX windows interface: Flexible strategies for multiple sequence alignment aided by quality analysis tools. _Nucleic Acids Res._ **24**, 4876–4882.

Thomson, G., and Klitz, W. (1987). Disequilibrium pattern analysis. I. Theory. _Genetics_ **116**, 623–632.

Tomaru, M., Matsubayashi, H., and Oguma, Y. (1998). Effects of courtship song in interspecific crosses among the species of the _Drosophila auraria_ complex (Diptera : Drosophilidae). _J. Insect Behav._ **11**, 383–398.

Tomaru, M., Doi, M., Higuchi, H., and Oguma, Y. (2000). Courtship song recognition in the _Drosophila melanogaster_ complex: Heterospecific songs make females receptive in _D. melanogaster_, but not in _D. sechellia_. _Evolution_ **54**, 1286–1294.

Tsaur, S. C., and Wu, C. I. (1997). Positive selection and the molecular evolution of a gene of male reproduction, _Acp26Aa_ of _Drosophila_. _Mol. Biol. Evol._ **14**, 544–549.

Tsaur, S. C., Ting, C. T., and Wu, C. I. (1998). Positive selection driving the evolution of a gene of male reproduction, _Acp26Aa_, of _Drosophila_: II. Divergence versus polymorphism. _Mol. Biol. Evol._ **15**, 1040–1046.

Tsaur, S. C., Ting, C. T., and Wu, C. I. (2001). Sex in _Drosophila_ mauritiana: A very high level of amino acid polymorphism in a male reproductive protein gene, _Acp26Aa_. _Mol. Biol. Evol._ **18**, 22–26.

Villella, A., Gailey, D. A., Berwald, B., Ohshima, S., Barnes, P. T., and Hall, J. C. (1997). Extended reproductive roles of the _fruitless_ gene in _Drosophila melanogaster_ revealed by behavioral analysis of new _fru_ mutants. _Genetics_ **147**, 1107–1130.

Wang, R. L., and Hey, J. (1996). The speciation history of _Drosophila pseudoobscura_ and close relatives: Inferences from DNA sequence variation at the _period_ locus. _Genetics_ **144**, 1113–1126.

Warman, G. R., Newcomb, R. D., Lewis, R. D., and Evans, C. W. (2000). Analysis of the circadian clock gene _period_ in the sheep blow fly _Lucilia cuprina_. _Genet. Res._ **75**, 257–267.

Weatherbee, S. D., Nijhout, H. F., Grunert, L. W., Halder, G., Galant, R., Selegue, J., and Carroll, S. (1999). _Ultrabithorax_ function in butterfly wings and the evolution of insect wing patterns. _Curr. Biol._ **9**, 109–115.

Wheeler, D. A., Fields, W. L., and Hall, J. C. (1988). Spectral analysis of _Drosophila_ courtship songs: _D. melanogaster_, _D. simulans_, their interspecific hybrid. _Behav. Genet._ **18**, 675–703.

Wheeler, D. A., Kulkarni, S. J., Gailey, D. A., and Hall, J. C. (1989). Spectral analysis of courtship songs in behavioral mutants of _Drosophila melanogaster_. _Behav. Genet._ **19**, 503–528.

Wheeler, D. A., Kyriacou, C. P., Greenacre, M. L., Yu, Q., Rutilia, J. E., Rosbash, M., and Hall, J. C. (1991). Molecular transfer of a species-specific behavior from _Drosophila simulans_ to _Drosophila melanogaster_. _Science_ **251**, 1082–1085.

Wicker-Thomas, C., Henriet, C., and Dallerac, R. (1997). Partial characterization of a fatty acid desaturase gene in _Drosophila melanogaster_. _Insect Biochem. Mol. Biol._ **27**, 963–972.

Wolfner, M. F. (1997). Tokens of love: Functions and regulation of _Drosophila_ male accessory gland products. _Insect Biochem. Molec. Biol._ **27**, 179–192.

Wu, C. F., and Ganetzky, B. (1992). Neurogenetic studies of ion channels in _Drosophila_. _Ion Channels_ **3**, 261–314.

Yamamoto, D., Jallon, J. M., and Komatsu, A. (1997). Genetic dissection of sexual behavior in _Drosophila melanogaster_. _Annu. Rev. Entomol._ **42**, 551–558.

Young, M. W. (2000). Life's 24-hour clock: Molecular control of circadian rhythms in animal cells. *Trends Biochem. Sci.* **25,** 601–605.

Yu, Q., Colot, H. V., Kyriacou, C. P., Hall, J. C., and Rosbash, M. (1987). Behaviour modification by in vitro mutagenesis of a variable region within the *period* gene of *Drosophila*. *Nature* **326,** 765–769.

Zehring, W. A., Wheeler, D. A., Reddy, P., Konopka, R. J., Kyriacou, C. P., Rosbash, M., and Hall, J. C. (1984). P-element transformation with *period* locus DNA restores rhythmicity to mutant, arrhythmic *Drosophila melanogaster*. *Cell* **39,** 369–376.

Zheng, X. Z., Zhang, Y. P., Zhu, D. L., and Geng, Z. C. (1999). The *period* gene: High conservation of the region coding for Thr-Gly dipeptides in the *Drosophila nasuta* species subgroup. *J. Mol. Evol.* **49,** 406–410.

Zheng, B., Albrecht, U., Kaasik, K., Sage, M., Lu, W., Vaishnav, S., Li, Q., Sun, Z. S., Eichele, G., Bradley, A., and Lee, C. C. (2001). Nonredundant roles of the mPer1 and mPer2 genes in the mammalian circadian clock. *Cell* **105,** 683–694.

Zylka, M. J., Shearman, L. P., Weaver, D. R., and Reppert, S. M. (1998). Three period homologs in mammals: Differential light responses in the suprachiasmatic circadian clock and oscillating transcripts outside of brain. *Neuron* **20,** 1103–1110.

Index